M000196117

PERSONAL EMERGENCY COMMUNICATIONS

Staying in Touch Post-Disaster: Technology, Gear and Planning

by Andrew Baze

Max Publications

TABLE OF CONTENTS

INTRODUCTION

How will you contact anyone if your landline phone, cell phone and Internet connection don't work? Will you be able to talk with family and friends after a serious emergency or disaster? Do you have a communications section in your personal or family emergency plan? Have you tested it? Do you even have a family emergency plan? If you don't have good answers to all of the questions above, don't worry. You're not alone. Most people don't.

The good news is that, since you're reading this book, you are not like most people. I'm honored to help you take your emergency planning to a new, improved level as you plan for realistic backup communications. With this book, creating an emergency communications plan will be easy, even if you currently know nothing about this topic. "A plan?" you say? "I'm just interested in the cool gear!" That's no problem, because I also cover gear. But let me ask you this: if you don't have even a simple communications plan, how will you be able to talk with family, friends and loved ones in a real emergency situation? How will you do something as simple as listening to the news or other emergency alerts if the power is out, your phones don't work and the Internet isn't available? This book will help you ask the right questions and come up with answers that fit you and your unique situation.

If you follow the guidelines outlined here, you will have the necessary tools to communicate in a disaster while most of those around you look at each other in shock, wondering who will help them get a message to their loved ones.

Andrew Baze

DETERMINE WHAT YOU NEED

In this section, we'll discuss a topic overlooked by most writers and others interested in radio or emergency communications: creating a plan that fits your situation.

Note: "emergency communications" are often simply referred to as "**emcomm**," a term I will use throughout this book. I confess that up until a few years ago, I didn't have a clear idea of what I would need in order to communicate with family, friends, neighbors or anyone else during an emergency. Writing this book certainly helped me form a plan, and reading it will help you in the same way.

The chapters in this section provide a planning overview. If you are not inclined to read the whole section, at least read the planning overview. I guarantee it will help determine where (as outlined in the upcoming sections of this book) you should best allocate your precious time, money and energy. In these first few chapters, I will cover:

- Who needs this information. (Hint: more than just you!)
- How a simple communications plan can make a world of difference.
- How to put a simple plan together. (With some freebies to help!)

Let's talk about your needs…

CHAPTER 1:

WHO NEEDS EMERGENCY COMMUNICATIONS?

This is a true story. It was winter 2001 and I had just run down three flights of stairs and paused outside my Redmond office building. A few moments earlier, Seattle and its suburbs were rocked by a serious earthquake. As I stood in shock with everyone else, I watched as many desperately tried to make cell phone calls with no luck. I remember one woman standing on the grass a few yards away, sobbing because she was worried sick about her twin daughters, who were in day-care. She wasn't alone; everyone there was worried about someone. I tried calling family, too, getting nothing but an annoying "beep beep beep."

Nobody's phones worked, although it wasn't because the tower in that area was significantly damaged. Even though the towers weathered the quake well, they were simply overloaded with traffic. Several thousand people evacuated buildings in the area and immediately attempted to make phone calls, but the cell phone networks couldn't come close to handling that amount of traffic.

In other areas, however, cell coverage disappeared because of power loss or tower damage. Being caught in a large earthquake was frightening enough on its own, but not being able to determine whether one's family was safe added to the fear and stress.

The Nisqually earthquake and the ensuing, temporary loss of traditional communications is just one example of how our most commonly used methods of communication can disappear without warning. A winter storm, hurricane or tornado, or another source of power outage could turn your phone into a useless chunk of electronic circuitry. And there's another possibility: someone could decide to shut towers off. On August 11, 2011, officials in San Francisco decided to shut off cell service on public transit in order to prevent a possible demonstration.[1] While this is an uncommon situation, it should still give you food for thought.

There are many real life, emergency scenarios out there, and you may remember a time when your phone didn't work. History shows us that one of the common elements of a serious emergency or disaster is a communications disruption or complete breakdown, at least in areas local to the event. Earthquakes, hurricanes, tornadoes, and even simple ice storms can break communications infrastructure.

Imagine that a natural disaster has just occurred in your area. After you ensure that you and your family are safe with shelter, food and water, and that no neighbors are in need of immediate assistance, what's next? You'll need to expand your area of influence to help even more people who are farther away, especially family or friends in the area. This is one of the key scenarios you need to address. And if there aren't any other locals who need your help, at least you can let loved ones outside the disaster-affected area know you and yours are safe.

Think about your unique needs. Do you have an elderly relative across town who you need to check on regularly? Do you work a significant distance away from your family? Keep thinking. Odds are good that you can imagine a situation where you would really need to talk with someone, *especially* if things were so bad that phones didn't work, your power was off and you lost Internet service.

---

[1] See news.cnet.com/8301-27080_3-20091822-245/s.f-subway-muzzles-cell-service-during-protest/. From the August 12, 2011 article: "The operators of the Bay Area Rapid Transit subway system temporarily shut down cell service last night in four downtown San Francisco stations to interfere with a protest over a shooting by a BART police officer, a spokesman for the system said today.

'BART staff or contractors shut down power to the nodes and alerted the cell carriers,' James Allison, deputy chief communications officer for BART, told CNET. The move was 'one of many tactics to ensure the safety of everyone on the platform,' he said in an initial statement provided to CNET..."

If you can't think of anything urgent (maybe you are single and currently living far away from family), consider whether you might want to help others who can't make an important call. Could you help your friends or neighbors? The scenarios I just mentioned and the ones you thought up are the context for the planning and equipment recommendations I will make throughout this book.

Now, let's briefly review options. When we're talking about technology options for emergency communication, we're talking about different types of radios. Because cell phones are wireless, they are essentially radios (although nowadays the computer portions of these radios grow more and more powerful). However, they are heavily dependent on existing tower and switching/routing infrastructure in order to function as radios. With a few exceptions, the radios we'll review are far simpler and independent of additional, local infrastructure.

Here is the basic list of technology we'll review, along with definitions, pros, cons and recommendations:

- Simple AM/FM/other listening suggestions
- FRS/GMRS radios (the bubble-packed, two-way radios available at most sporting goods stores)
- CB radio
- eXRS radio
- UHF/VHF amateur radio
- HF radios with unique antennas that allow regional communications (up to ~400 miles)
- Satellite phones
- Personal locator beacons and related devices
- Scanners
- Basic backup power supply recommendations

Please note that I sometimes make specific product recommendations, for example, when I discuss scanners and secure communications. Nobody paid me to endorse any product. In some cases, the products I describe are so unique that they currently have no serious competitors. Ideally, customer demand will result in more variety and competition in the marketplace.

You've seen a brief description of what I'll be discussing. I will not be discussing marine, aviation, and MURS radios. All of these radios are generally similar to FRS/GMRS because they operate on a fixed set of frequencies. You get to pick a channel, and if it's busy, you have competition and may not be able to get your message across. The frequencies for these different radios are in different "bands" (ranges on the radio frequency spectrum) and can be used at longer or shorter distances. Additionally, in many areas, these radios are less common than FRS/GMRS radios. If you need, have or want a MURS, marine, or aviation radio, you will still get value from this book, because the way you use these radios is very similar to the way you use FRS/GMRS radios.

I also won't be discussing digital modes with amateur radios (with one classic exception), a topic that usually fills up an entire book all by itself. If you are a licensed ham or are inspired to become one after reading this book, I recommend that you do some basic research on digital modes, because they can add more flexibility to your emergency communications plan.

Speaking of planning, let's look at how you can very easily create a simple plan that will take your emergency preparedness to a whole new level!

CHAPTER 2:

CREATING A PERSONAL EMERGENCY COMMUNICATION PLAN – THE BASICS

Travis stood, cold and frustrated, outside his office building in the middle of the city. After the earthquake, it had taken an excruciating 25 minutes to get from the 12th floor to ground level in the stampede of office workers. The people from the 20th floor probably still had another hour before they'd make it out. People milled about him looking frightened and cold, many having forgotten to grab their jackets.

Travis remembered his, but he still shivered in the cold, autumn air. A far deeper chill shot through him as he wondered again if his daughter Emily was safe. As part of his routine, Travis dropped Emily off at day-care earlier that day, just like on any other day. It was about six miles away, between work and home.

While he knew the day-care building wasn't very old and the employees were trained on emergency procedures (as they were all required to be in his state), he wanted more than anything to know that his little girl was safe. He checked his phone to see whether the school had tried to contact him. No texts. No email. No voicemail. And no signal.

Travis stamped his feet against the cold. His car was parked in the lot across the street, but, looking at the gridlock that was forming quickly, he knew that it would take hours to drive the short distance to the day-care. To add to the uncertainty, he would have to cross a bridge that spanned the nearby highway, assuming the bridge was still in one piece and structurally sound.

> *He headed to his car to get warm and check the radio. Halfway there he realized he had left his keys in his desk drawer. Luckily, he had a spare in a hide-a-key attached to the car frame. He quickly found it, unlocked the door, got in, and turned on the radio. A tense announcer discussed the obvious, that they had just experienced a serious earthquake, the roads were clogged, some older buildings had collapsed, and emergency services were not available in many areas. The governor had declared a state of emergency, and FEMA was mobilizing, due to arrive with aid at some point. But when, Travis wondered? How could he find out if Emily was all right?*
>
> *All was not lost. Travis had given some thought to preparing for a large-scale emergency, and had a small bug-out-bag[2] in the trunk of his car, as well as an extra jacket. He was more prepared than most, but this planning couldn't help him determine whether his daughter was safe.*
>
> *After listening to the radio for a few more minutes, Travis made up his mind. He got out, donned his jacket and backpack, took a last drink from the large water bottle in his car (he had another smaller bottle in his bag), locked up, and started walking toward the day-care. More than anything, he wished he could talk to someone there. Would anyone even be there by the time he arrived?*

How will you be able to communicate in an emergency situation? Will you be able to talk with anyone other than those people within shouting distance? Can they talk to you? If you were in Travis's shoes, what options would you have?

If you can't make it home and you know your family is waiting and worried about you, will you be able to talk to them somehow? Or, what if you have close friends in an area just hit by a tornado – how will you

---

[2] A Bug-Out-Bag, aka "BOB," is the portable set of gear you should have handy in each vehicle. It may also be in the form of a "three-day bag" or "emergency kit." See more details in Chapter 17.

check to see if they're all right? Will you have to wait for a notification from the Red Cross? Or will you be able to figure it out on your own? Many businesses have a "business continuity plan" or "emergency response plan," to ensure the business doesn't completely fail after an emergency. City, county, and state governments have emergency management offices, with employees whose full-time job is to plan for disasters. To make sure your family and friends are safe after a disaster, you should also have a personal plan. You may not have an emergency planning office at your disposal, but you are smart enough to read this book, so you are well on your way.

Let's take it a step further. How could you communicate with your neighbors – the elderly widow down the block who always waves and says hello, or the young couple with three kids across the street? Is there any way to talk with them if there is a blizzard or tropical storm raging, other than trekking out in it yourself?

The odds are slim that you'll be able to accomplish any of these tasks without a basic emergency communications plan. But what's the best way to figure out what to do? Of course, it depends on what you need. Let's start with the basics.

Step 1: Grab a piece of paper and a pen, and answer these questions. *Don't skip this part.* If you don't think about these questions and come up with good answers, your plan could have critical gaps.

Note: I strongly recommend you write down your answers. It'll only take a moment, and you'll be rewarded for this small time investment!

Key Questions

1. Who are you?

At first glance, it probably sounds like a silly question, but think about this for a minute. If you are a man responsible for a wife and three young children versus a single woman hundreds of miles from home, you may have very different needs. Are you in the military? Do you travel frequently?

Here's another way to look at the question: *who depends on you?* Are you responsible for your family's safety? Do you have a feeling of responsibility to take care of your friends or neighbors? Do you need to take care of a group of people who work for you?

Think about who you are in the context of your various roles in life. These answers will take you to the next question…

2. With whom do you need to communicate?

Go back to question number one. Do you need to reach family members close by? Maybe you have very close friends in the next state. Maybe you need to help take care of neighbors down the block or across the valley. Or maybe you are part of an organized (loosely or tightly) group of preparedness-minded friends, classmates from your CERT course, or others.

3. Are they able to hear you and communicate back with you?

This may seem like another dumb question, but it is critical. Let's assume you set up a fancy radio station that can reach your parents in the neighboring state. If they don't have a radio with which they can hear you, then you don't have a useful solution. They need to be able to respond. You may have to minimally equip and train some people as part of your plan. But don't worry. I will show you how to do this in multiple ways, with little effort!

4. How far do you need to communicate?

Are your friends or family close by, thousands of miles away, or somewhere in between? When we look at your options later in the book, this will also be important. And again, don't worry. Communicating at great distances is quite do-able. It just requires different equipment. Think about distance in these categories:

- **Very close** – same neighborhood
- **Close** – same town/city
- **Regional** – same state or part of a state
- **Long distance** – a different state, province, region, country, or continent

5. How often will you need to communicate?

Do you need to talk with someone more than once a day? During which time(s) of day do you need to speak? Is there a particular time that works best? Do you have to consider different time zones? In Chapters 3 and 4, I'll show you how you can set up a schedule, known as a "calling clock," which will enable you to reach someone more reliably and save precious power at the same time.

6. Do you need to be mobile?

Do you expect to be staying at home? Do you expect to have your car, truck, or SUV available? Do you expect to be on foot, carrying everything you need in a backpack? There is a big mobility difference between a desktop model radio running off a deep-cycle marine battery and a hand-held radio running on AA batteries.

7. Will you need to transmit data?

In some scenarios, emergency radio teams are prepared to transmit lists of supplies and needed medical equipment to other teams or government agencies as part of disaster relief efforts. This may not apply to you personally, but you should still think about it. This can be done with as little as a handheld radio and an inexpensive netbook computer. Note: while this book doesn't discuss data transmission technology, you should still identify whether this is a need and do more research.[3]

8. Will you have a power supply?

What kind of power will you need? Unless you're using smoke signals, you will need a power source to run your communication device(s). Depending on each device's power consumption and how often you use it, you may need a generator, several AA batteries, or one of the many other options. Don't neglect this area. All of your planning and preparations will be useless if you don't have the power to run your equipment. I'll show you more power options in Chapter 15.

---

[3] One good information source for digital communications modes in amateur radio is the ARRL's "VHF Digital Handbook." It covers both voice and data digital transmissions technology. If you are interested in transmitting data over HF frequencies as well, consider the companion: "HF Digital Handbook."

9. Do you have the skills and equipment you need?

Unfortunately, this part of emergency communication planning is often overlooked (especially the skills part). Answers to the previous questions will give you an idea what you'll need. If you haven't answered them, don't waste your time and money by buying a bunch of new gear. You won't know what to get until you invest a little thought. Once you have a clear vision of what you need to accomplish, then you will be able to identify gaps in equipment and skills. NOTE: equipment is usually easy to acquire, assuming you have a few dollars available, and often simply involves making a purchase online. The **skills** needed to use the equipment effectively are more challenging, and you'll need some practice. Practicing with your gear needs to be part of your initial plan and part of the ongoing maintenance of your plan.

How are these questions helpful? Don't underestimate the value of a good question. In fact, I'll wager that most people have never answered the simple question, "How will you respond in an emergency?"

Sometimes asking a simple question can cause a lot of positive action to happen. (More than once in my life, a very simple question has caused me to say, "Huh?" and significantly change some of my behavior!) Please take a serious look at the questions above, and as I mentioned before, please **write down** your answers. As I discuss later, you should also review those answers with your spouse and any other key members of your plan.

Congratulations! You are already ahead of the game. And you have extended your lead further by writing down your answers. Writing is much more useful than simply producing fleeting thoughts and then going about your business, especially considering that many of these thoughts often fade away. Once you write something down, you move it from a "thought" space to a "real" space, and this is a good way to get started. Now that you have written answers, you are ready to move on to the next chapter.

CHAPTER 3:

EMERGENCY COMMUNICATION PLAN PART II – YOUR CALLING CLOCK

Paula sat staring at the radio on her husband's desk and wondered what to do next. Ronald had a fancy HF ham radio and had even showed her how to use it. He always carried a radio with him when he travelled (he was a radio enthusiast — OK, a fanatic, they both knew it), and she wanted desperately to make contact, to let him know that she was all right. The tornado had torn up several homes a couple of blocks away, and damaged the phone and power lines. The lights were out and her laptop couldn't connect to the Internet, but she and the kids were safe and warm.

How could she let her husband know they were safe? Paula didn't want him to panic and try to take an unnecessary flight home from his job site six hundred miles away.

Paula turned on the battery backup power supply and then powered up the radio. It was already set to the frequency Ronald would use to call her. She noticed the blinking yellow light. The charge was low. She remembered that as soon as it turned red, she would only have a few minutes of power left. What should she do? Should she listen until it ran out? Should she only listen for his call in the evening? Should she try to call him herself? She knew calling used far more power than listening and would drain the remaining charge even further. If she only knew when Ronald would be calling or listening on the other end…

In the previous chapter, you answered basic questions about with whom you need to talk and the distances you would need to cover. These answers comprise the cornerstones of your emergency communications plan.

I'll discuss specific equipment capabilities and options later, but in the meantime you need to flesh out your plan a little further. To make the plan realistic, the people you will need to communicate with also need to know what the plan is. How can you do that?

The type of plan I'll show you in this chapter will tie your previous answers together and integrate the people, technology, and schedule. It's called a calling clock, which is a schedule of times, people (stations), and frequencies or other relevant numbers (e.g., phone numbers), which should be known to all of the participants of the plan.

Here is an example of a very simple emcomm plan:

In the event of an emergency, your brother will call you at 6:00 PM Eastern Time on your cell phone. If he hasn't called by 6:05 PM, you call him.

That's it. But it seems a little too simple, doesn't it? Let's start looking at it critically. Do you know the phone numbers of everyone you care about? I've been using a smartphone with its built-in address book for a long time, during which many of my family and friends have changed their phone numbers many times. Now I currently remember exactly **two** phone numbers. The first is my parents' (because it hasn't changed since I was a little boy) and the second is my wife's (because I forced myself to remember it, even though she's on speed dial). That's right, I remember only two phone numbers. That is not going to be a big help. I have more family and friends I will want to talk with. Although you may remember more phone numbers or have a much better memory than me, I suggest you write them all down anyway. Don't just store them in your phone, because if the charge dies or you lose the phone, you will have lost critical contact information. Paper can be more useful than electronics in some ways, and this is one of them!

Let's get back to the emcomm plan. It's too simple because it is missing some key data. We will go one step further by adding more specific information:

In the event of an emergency, your brother will call you at 6:00 PM Eastern Time on your cell phone. Your number: 426-555-1212. Brother's number: 426-555-2121. He always calls you first. If he hasn't called by 6:05, you call him. If one of you can't reach the other, you leave a voicemail message. If contact is not made within four hours, one of you will leave a message for the other with the parents [out of the area], at 426-555-1122.

Not only did we add specific phone numbers, but we also added a Plan C and Plan D. Plan B indicates that if your brother does not call, you try calling him. Plan C is that if one of you or the other does not answer, a voicemail is left. Plan D is that if the call is not returned within a set time, a subsequent call is made to an out-of-area contact, in this case the parents.

Does this seem like overkill? Remember, we are talking about an **emergency,** not routine communication, or a weekly chat with Mom. In an emergency, a little planning now, especially something as simple as documenting times or critical phone numbers, will go a long way toward reducing the tension, confusion, and possibly even the danger of the situation. And if writing all of this down looks like a lot of work, don't worry. It is not. I will make it easy for you! Keep reading.

Let's review the key components of our emcomm plan so far:

1. Who: You and your brother
2. When: Daily at 6:00 P.M.
   a. Backup plan – call parents after four hours with no contact
3. Where: Wherever you are
4. Equipment or frequency: cell phones, phone numbers for everyone
5. Backup plan: You call at 6:05 P.M., leave message; if no contact, leave message with parents

Is this an adequate emergency communications plan in the event of a natural disaster? Probably not, because it's quite possible that people will have no cellular reception in such a situation. But it is a plan, and it contains a schedule.

In a moment, I'll take it a step further and discuss a much more serious calling clock, incorporating some additional communications capabilities. And to make it easier to organize all the information, I'll use a template, the same one that is available to you here:

www.EmergencyCommunicationsBlog.com.[4]

You can see in the earlier example why answering the basic questions around who, when, and how is very important. If you start asking, "What if the power runs out on my phone?" or "What if that frequency is busy?" you will probably end up finding more and more questions pertinent to potential emergency situations. This plan is the perfect tool for documenting your answers and other key details you'll need in an emergency.

How to Use Your Calling Clock/Plan Template

Here are the major sections of the template, and the type of information that goes into each one.

Method: Phone, radio (and type), satellite phone, etc.

Who: The key players: you, your spouse, etc. Note: You may need a variation of the plan for different people, and each variation should be documented. Once the initial plan is in place, subsequent variations will mostly be a "cut and paste" operation with a few modifications to fit each situation.

When: Right away, then every hour? How long will you try each hour? How often to call or how long to monitor (if using radio)?

Where: This may matter if you are contacting people out of area or in a different time zone (which should also be specified in "When"). If the people in your plan are all in the same area/time zone, you may want to omit this section.

Phone number(s): Area codes and phone numbers for all participants.

---

[4] An Emergency Communication Plan template and other resources, ideas and articles are available at www.EmergencyCommunicationsBlog.com.

Frequency: All frequencies or channels used in this part of your plan. If you are using a repeater, make sure you include any tone and offset information.[5] If one or more members of the plan do not know how to modify any settings on the radio, that is OK. Someone else still may be able to make the changes for them, if needed. Or they may even be able to figure it out for themselves, if they have a manual available.

Notes: Any other important details, especially around power management, e.g., "If the phone/radio reaches half/quarter power, only listen or call on even hours [versus every hour]." If there is no answer, then what?

Other details section: This is a good place to put a reminder about using other forms of communication or listening.

Figure 3-1 (next pages) is a more detailed example of a calling clock and overall communication plan. It uses the following technologies:

---

[5] Tone and offset: If you want to talk on a repeater, you will probably need to set or modify the default tone and offset that the repeater requires.

Many repeaters listen for a special tone, which is included in the signal that carries your voice when you transmit your spoken words. Usually this is programmed into your radio when you set up the frequency to use with the repeater. Without that special tone, the repeater will not repeat what you transmitted. Using a tone this way is called "tone squelch Note: Some people or radios may try to confuse you with technical jargon by calling this "CTCSS" or "PL" – don't let that throw you. These letters are simply referring to the tone you use.

The offset tells your radio what distance to move up or down the frequency spectrum, in order to match what the repeater will receive and transmit. Here's the key thing to remember – repeaters listen on one frequency and re-transmit on another. For example, a repeater will receive a signal on 146.050 megahertz (MHz), and then re-transmit that same content on 146.650 MHz. This means that when you use your radio, you will press the transmit button, your radio will transmit your voice on 146.050 MHz, the repeater will receive your transmission, and then it will re-transmit it on 146.650 MHz. The offset is simply the difference (and the direction, up or down) between the two frequencies. Offset is usually standardized at 600 kHz for the 2M band and 5.0 MHz, either upward or downward (shown as "+" or "−") in repeater setup information.

Once you make sure these two settings match the repeater you want to use, you're set!

Don't forget to refer to your radio's manual, and make sure you know how to save a frequency and repeater settings to memory.

Plan A	Method	Cell phone
	Who	Chris and Helen
	When	As soon as possible, after emergency happens
	Where	Same time zone
	Phone number	Chris cell: 436-555-1212, Helen cell: 426-555-2121
	Frequency	NA
	Notes	If no answer, leave a voicemail, followed by text message. If no answer in one hour, proceed to Plan B. If phone is less than 50% charged, shut it off and turn it back on for 15 minutes, every five minutes before the hour until ten minutes after the hour, every hour. If no signal is available after the first five minutes, turn the phone off and try again the next hour.
Plan B	Method	Cell phone and landline phone
	Who	Out of area contacts, Jack and Kimberly
	When	Every hour, on the hour, one call
	Where	Same time zone
	Phone number	Jack's cell: 426-555-1212, Kimberly's land-line: 426-555-2121
	Frequency	NA
	Notes	If they answer, leave a message with details, including any alternate contact information for yourself. If no answer, leave a voicemail, followed by text message to Jack.

FIGURE 3-1: Example emergency communication plan.

Plan C	Method	FRS/GMRS radio
	Who	Chris and Helen
	When	Every hour for 15 minutes, starting on the hour
	Where	Same time zone
	Phone number	NA
	Frequency	Channel 17, privacy code 10 (the default setting for all of our radios)
	Notes	Try calling for 15 minutes every hour, calling every 2–3 minutes.

[Note: if you have enough backup batteries, other information may be learned by setting the radio to "scan" during the rest of the hour, and listening to nearby conversations.] If radio(s) are less than 50% charged or batteries are scarce, call only on even hours. Move to plans D and E. |
Plan D	Method	Radio-VHF repeater
	Who	Chris and Helen
	When	Every hour for fifteen minutes, starting on the hour
	Where	Same time zone
	Phone number	NA
	Frequency	146.440 MHz (station 1 in memory, the KE7ZZZ repeater, tone +103.5)

FIGURE 3-1: *(continued)*

Plan D (cont.)	Notes	As with the phone schedule in Plan B, 15 minutes every hour, calling every 2–3 minutes. Chris call sign: KX7AB8, Helen call sign: KX7AB9. If the repeater is too busy, not working (no beep-beep is heard after your transmission), or has been taken over by emergency services for government-only use, go to Plan D. [Note: other information may be learned by simply listening on this repeater, if sufficient power is available.] If radio(s) are less than 50% charged, call only on even hours.
Plan E	Method	Radio - VHF/UHF
	Who	Chris and Helen
	When	Every hour, for 15 minutes, calling every three minutes
	Where	Same time zone
	Phone number	NA
	Frequency	147.010 MHz (simplex, no repeater) for seven minutes, switching to 440.220 for the next seven minutes.
	Notes	If radios are less than 50% charged, only call on even hours. If you encounter someone else on either of these frequencies, you can ask them to rebroadcast your transmission, especially if they have more power or are located farther away. This approach will significantly increase your transmission distance.
Other details	Scanner	Turn on scanner after emergency
	NOAA radio	Monitor NOAA alerts, as well as AM or FM news stations, if they can be received.

FIGURE 3-1: *(continued)*

1) cell phones, 2) landline phones, 3) FRS/GMRS radios, and 4) hand-held UHF/VHF amateur radios. You may not have all of this equipment available to you (and if not, I will show you what you need), or you may have more. In either case, it is just an example.

As you can see, this is a much more thorough plan, but at the same time, many of you (especially the very detail-oriented, contingency planners) will think of other questions that haven't been addressed by this plan. You will need to make your own and make sure it answers the most important questions for your unique situation.

Take the template I've provided, or modify one of the examples and develop your own plan. If you also want to add equipment, go ahead and add those sections to the plan now. Since you've written down how you will eventually use that equipment, it's even more likely you will configure your new gear and test it realistically once you get it. For now, just write "In progress" in your plan for that specific gear, to ensure nobody else who needs to use your plan gets confused.

If you have any questions about my template, your own plan, or would like a personal review of your emergency communication plan, please feel free to contact me by posting your inquiry on www.EmergencyCommunicationsBlog.com.

CHAPTER 4:

YOUR COMMUNICATION PLAN, PART III – EQUIPMENT OPTIONS

Did the sample calling clock and communication plan from the previous chapter whet your appetite? Have you written up your own plan and identified some important gaps you need to fill right away? If so, keep up the good work! If not, go do it!

You can see from the sample plan how the addition of several options (e.g., different radios) can significantly increase your ability to communicate if phones stop working. However, you may be asking yourself, which equipment is capable of what? How much does each radio cost? Is a license required to operate a certain radio? Now that you have answers to the key questions (who you will be talking with, when, etc.), and more good questions about gear, let's dig deeper.

There are a few variables we must consider when adding capabilities to your emcomm plan. What do you need your equipment to do? How much money can you spend? Can you devote a little time to study for a license exam?

You'll shortly see that if you want to do zero training or studying, you have good, inexpensive, short range, two-way options available. But if you want to go long range without getting any training, you'll need to spend quite a bit more money to invest in a satellite phone system. And not only is the initial cost to purchase a satellite phone high, you will also need to pay monthly service fees, even if you never use the phone.

If you have a limited budget, as most of us do, and are willing to learn a little bit about how radios work, you have a much less expensive option: amateur radio.

Let's step back for a moment and look at some basics of wireless communication in an emergency situation. There is a variety of important equipment to consider, and I'll walk you through each of the options.

For now, you can review the following table (Figure 4-1). It lays out some basic considerations and will give you more context when you read about those options, their pros and cons, and my specific recommendations.

On the left, you can see gear options reviewed in this book. (Note: FRS/GMRS options also cover MURS, Marine & Aviation radios.) In the top row are criteria for choosing technology, specifically the range, training needed, and price.

Gear and Needs

Gear / Needs	Short range	Med. or Long range	No tng.	Some tng.	$	$$	$$$
AM/FM/NOAA radio/Scanner			X		X	X	
FRS/GMRS radio	X		X		X		
CB radio	X		X		X		
eXRS radio	X		X		X		
Satellite phone	X	X	X			X	X
Amateur UHF/VHF radio	X			X	X		
Amateur HF radio	X	X		X		X	
Personal Locator Beacons	X	X	X			X	

FIGURE 4-1: Gear compared to cost, range, and training (tng.) needed.

LISTENING STRATEGIES

In the age of iPods, smartphones, tablets, Internet phones and other high-tech wizardry, people get more and more of their news from the Internet. A survey by IBM in 2007 indicated that Americans in certain demographics routinely spend more time in front of their computers than their televisions. The trend has continued upward since then. In 2010, a Forrester survey indicated that Americans in general spend the same amount of time on the Internet as they do watching television. The interesting part: they are spending the same amount of time watching television, but they are eliminating newspaper and radio as information sources, as you can see in the following chart.

SECTION II CHART: How do we get our news? From *Pew Research Center for the People & the Press*

Where Do You Get Most of your News About National and International Issues?

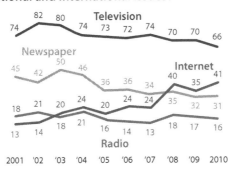

PEW RESEARCH CENTER Dec 1–5, 2010. Figures add to more than 100% because respondends could volunteer up to two main sources. If asked more than once in a calendar year, trend shows final datapoint from each year.

What does this mean? One simple conclusion is that people are more and more reliant on the Internet for their information. Before long, more people will go to the Internet than to television for news.

Does this matter? Not much in an emcomm context, because during a disaster, you probably won't have access to either one!

However, after a disaster, you'll probably still have one real-time information source available: radio. But there's a small hitch. People are less and less likely to have one that will work if the power is out. This is an easy problem to solve, but we can do more than just getting a simple AM/FM radio that runs on AA batteries.

In the next chapters I'll discuss gathering critical information (news and more) and I'll show you some more effective and useful solutions.

CHAPTER 5:

AM, FM AND BEYOND

"Press 'play', Mom!" said Jack, pointing at the television screen and inadvertently spitting out little bits of popcorn from his full mouth with each "p."

"Yeah, let's go!" his brother Josh echoed.

"Hold on a second," Susan replied. "And keep your food in your mouth!"

She wanted to catch the latest news on the tropical storm heading their way. Earlier in the day there had been speculation it could turn into a hurricane. With her husband Lance, three states away on a remote fishing trip with his college friends, Susan was especially concerned about making sure she would be able to get her family and their pets to safety if the weather got really bad.

She drummed her fingers, waiting for the commercials to finish before the weather report would continue. If things weren't going to get worse soon, she and the boys would probably be OK, and they could get on with the show.

The TV and the lights flickered. A second later, everything went dark. The power was out.

"Noooooo!" the boys wailed. They had been waiting for their Friday night movie all week long, and now it had been spoiled.

Susan had a different concern: the storm was obviously getting worse. Was it now headed right toward them? How would she know if there was an evacuation order? They had almost been forced to evacuate two years ago, and it was a chaotic situation, even with Lance there to help.

Regardless of the source of the outage, they would be fine in the short term. They had food, water, a new LED lantern with spare batteries, and the summer evening was cool, but not uncomfortably so.

"Wait here for a minute. Eat your popcorn and don't make a mess," Susan said, as she fumbled her way to the kitchen. She found and opened the tool drawer, pulled out a small flashlight, and turned it on. So far, so good. She crossed the kitchen toward the garage. Once inside, she made her way to the back, reached up to the top shelf and retrieved a box labeled "emergency radio." She set it down, opened the lid, and pulled out a radio and one of the four-packs of lithium AA batteries.

She took everything back to the living room, where Jack and Josh were munching on popcorn and giggling in the dark.

"OK, boys, let's see what's going in the world." Susan popped the back off the radio and inserted four AA batteries. It had a hand crank, but she wanted the news now, not in three minutes. And turning the crank, she reasoned, would give the boys something to do in a few minutes. She turned the radio on, switching to the local NOAA weather channel.

"Tropical storm Bart is holding steady. Winds of up to forty miles per hour are expected throughout the day, Saturday, starting in the late morning..." Well, at least it wasn't too much worse.

"OK, Jack, you get to crank the radio to charge the internal battery — at least one hundred cranks. Josh, you count. Get started. I'm going to get the laptop from the bedroom and we're going to have a special blackout movie party!" Susan would use her ear buds to continue monitoring the NOAA station while the boys watched their movie. They may not have power, but for the meantime they were informed and safe.

At the most basic level, half of communication is receiving a message, and since we're far removed from waiting for the village crier and his bell to alert us to the latest news, our best bet is radio.

Every family needs a radio powered by a couple of AA batteries or a small 9-volt battery. A good radio can run for many hours with low power, especially when using headphones, which typically use less power than using the speaker.

Nowadays, unless you already have a plain-Jane AM/FM radio and don't want another, you might as well get one that has a built in flashlight, NOAA weather alerts (a weather band), and a hand crank and solar panel for alternate power. Some even have USB ports or other proprietary power jacks, which you can use to charge other portable electronic devices as you turn the crank. This may seem like a lot to find in a single radio, but they are much more common today than ten years ago.

Let's quickly review the radio frequency bands. The FM (Frequency Modulation) band has most of the music stations, and the signals are usually relatively local.

AM (Amplitude Modulation) signals are usually local, too, but there is a special advantage to AM radio. In the evening or night, you can receive broadcast signals from far away. Why? Because at that time of day the sun's radiation is not ionizing a certain layer of the atmosphere, which would otherwise absorb these signals. At night they are able to rise higher before they bounce back to Earth, causing them to travel farther. Why does this matter? Because it means **you will probably be able to hear news about your area, even if the transmitters in your area don't work.** Do not think AM is just for listening to talk radio or "the oldies." It is also quite useful for regional and even longer-distance newsgathering.

This brings us to shortwave (SW). Many of the "emergency radios" on the market also include one or more shortwave bands. If you can receive shortwave radio, you will be able to listen to stations from all around the world. Shortwave radio is technically also AM radio, although the frequencies are higher than "broadcast AM" frequencies in the US. If you would like to know more about the frequencies, you can get more info here: www.SWLing.com.

As with amateur radio, certain bands (ranges of frequencies) are allocated for international broadcast. Shortwave broadcasts may also provide valuable information transmitted from other countries, near or far, if local radio stations aren't transmitting. Shortwave listening is an interesting hobby, and you can hear broadcasts from nearly any location, in any language, any time of day or night. It can certainly play a valuable information-gathering role in your emcomm plan.

Now I'll show you another interesting option that comes with many emergency radios: NOAA.

What Is NOAA and Why Do You Need It?

NOAA stands for the National Oceanic and Atmospheric Administration, and is part of the U.S. Department of Commerce. That may sound pretty boring, but there are many times when their messages are very interesting.

This technology is not just about the weather. NOAA is an "all hazards" network, which means they will also alert listeners about natural disasters, oil spills or chemical releases, and public safety issues, like a 911 outage. As you can imagine, this type of information would be very useful in an emergency. You need a weather radio, or a radio that receives the NOAA frequencies, as part of your emergency communication plan. Even though the communication will be one way (you listening), you will get critical information from this source.

How does it work? For extreme weather reports, or when other life threatening events happen, NOAA's National Weather Service makes special announcements on certain radio frequencies, using one or more of their thousand-plus radio stations across the country. According to their design, each station covers a small area because many announcements are often only pertinent for a specific area. NOAA also regularly broadcasts weather forecasts on these frequencies, which can be handy if you want to know the weather forecast immediately, especially when you don't have access to the Internet. In addition, I have found these weather updates to be more informative than many of the basic weather sites I've seen on the Internet. The announcements are very detailed and updated frequently.

How Does the Emergency Alert Feature Work?

Radios that receive only these NOAA frequencies are usually called weather radios. Many ham radios and some FRS/GMRS radios (the type of handheld, two-way radios you can get at the sporting goods store, which we will cover later) will also receive these frequencies. You can also get radios that are designed solely for listening to these stations. Some of these radios quietly listen for emergency announcements, and will alert the user when it hears them. Any good emergency radio will receive weather radio channels. Make sure you have a radio that gets NOAA channels.

One other thing you should be aware of with NOAA radios: some of them allow you to input a local code, called a SAME (Specific Area Message Encoding) code. This code ensures that your radio receives messages intended for your area. You can find your SAME code here: http://www.nws.noaa.gov/nwr/indexnw.htm.

Simple setup information should be included in your radio's instructions if it has this functionality.

You can find more information on NOAA at www.weather.gov/nwr.

Honorable Mention: TV and Crystal Radio

In the last couple of years, I've seen more and more small, portable, **battery-powered, flat screen TVs** available for increasingly lower prices. One of these could also come in handy. A picture can be worth a thousand words, and you may be able to get some detailed pictures with one of these devices, assuming your local broadcasters are up and running.

Warning: don't buy an old, used TV that doesn't receive digital TV signals (which is what the vast majority of broadcasters send, since 2009). You can buy a new one for as little as $50! There are many options available.

The distance you'll be able to receive TV signals will probably be similar to the ranges in which you'll be able to receive FM radio signals, which may limit your options in a serious regional disaster. Take a look around online and you'll be able to find a good one. Make sure you check the customer reviews.

And last, but certainly not least, we have the venerable and relatively obscure (especially to today's youth) **crystal radio**, originally developed in the late 1800s. Yes, this is actually antique technology. But here's the crazy part: it needs no battery or any other power source, other than the radio waves it receives. That's right – no other power source. And you can make one from just a few parts, or an inexpensive kit found online or in your local hobby shop. They mostly seem to be designed as kids' science projects, but if you want a radio solution that requires zero power, this is the one for you.

I did some searching for more commercial, inexpensive crystal radio kits, but didn't find anything that got many good reviews. Specifically, most of them didn't seem to work as promised. However, I know the technology works. In fact, I made one when I was a kid, and I remember listening to AM radio with it. I wish I still had that radio to play with. If you find a good model, please let me know!

Some fascinating options for learning about the history of crystal radio and how to build your own can be found here:

- bizarrelabs.com/foxhole.htm
- www.satcure-focus.com/hobby/page6.htm
- www.crystalradio.net

AM/FM/SW Pros & Cons

Pros:
- Broadcast radio stations often have very good backup power solutions, so they often continue to transmit during emergencies.
- With AM or SW radio, you can receive transmissions from outside your area.
- Radio listening, especially with headphones, consumes little power.
- Many radios are easily portable, fitting into your pocket or backpack.

- Radio will allow you to receive local government emergency alerts, NOAA alerts, or Emergency Broadcast System[6] alerts.

Cons:
- None! A high-quality portable AM/FM/Weather radio is a must.

Recommendations

If you don't have a portable AM/FM radio, get one right now. Your car radio doesn't count. You need one that you can stick in your backpack, set on your nightstand, or take with you anywhere and use without having to plug into AC power (your power-grid-connected wall outlet or a generator).

FIGURE 5-1: Freeplay Eyemax, a very durable and high quality emergency radio.

If you have nothing and are on a strict budget, go find an inexpensive AM/FM transistor radio that uses AA batteries at your local Target, Wal-Mart, or other discount store, and buy a block of AA batteries to power it. That's the quick, inexpensive, easy version. But you can do far better!

If you do a little research, you'll find that for the same price or a little more money, you can get a higher-quality radio that has alternate power supply methods, e.g., a built-in solar panel, a hand crank for recharging batteries, and built-in rechargeable and disposable battery options, as well as an AC adapter when grid power is available.

One of my favorite radio brands is Freeplay, and I have been very impressed with their quality. Their gear is sturdy. I've been using an Eyemax

---

[6] The Emergency Broadcast System was set up to allow the president to communicate with the population efficiently in the event of a serious, national emergency, using radio and television.

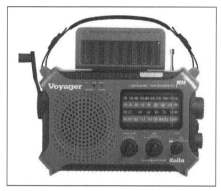

FIGURE 5-2: The Kaito KA-500 Voyager has many power options, including adjustable solar panel.

model that provides AM, FM, and Weather for several years now, and it still runs just fine. This model in Figure 5-1 (like everything I've seen from Freeplay) is very durable and ranges in price from $60–$75 at the time of writing.

One other good option I've used is the Kaito KA-500 Voyager. It has a few interesting features. In addition to the rechargeable battery pack, you can power it with disposable batteries. Also, the solar panel on the top can be tilted to face the sun at an optimal angle, which will improve your charging efficiency. It also comes with a hand crank. In addition to AM, FM, and NOAA weather bands, it also has two shortwave bands, which is a nice bonus.

There are many other hand crank radios on the market nowadays, but whichever you get, you will be better off if you do a little research first. They are not all of the same quality, and some are downright junk. If you're going to spend the extra money, you should consider high quality brands, such as Freeplay, Kaito, or Eaton.

CHAPTER 6:
ADVANCED LISTENING: SCANNERS

True story: Several inches of snow had fallen in the Seattle area, which was a lot for the region. Some of the snow melted and then refroze overnight when the temperature dropped, coating the streets in a treacherous layer of ice. The city was a mess, and dramatic, multi-car pile-ups were constantly shown on the local television news. I didn't absolutely need to go out, so I stayed at home, but when I looked and listened for any news about the road conditions in my neighborhood, I found nothing. Since my neighborhood didn't have many long, steep hills, it wasn't considered newsworthy. However, we still had hills and I was pretty sure it could be quite dangerous. I didn't have news, but when I turned on my scanner, I learned what was really going on.

Some roads were impassable because of parked, stuck or crashed vehicles. Some were clear, plowed, and safe. And some looked safe, but were not.

How did I know? The bus drivers told their dispatcher, each other, and indirectly, they told me. Since the bus radios used public frequencies (as the vast majority of all public services do), I listened on my scanner as they discussed the routes in my area.

Now I knew which routes to avoid if I decided to trek out.

Scanners are designed to receive transmissions across a wide range of frequencies, which aren't otherwise covered by ordinary radios. This can be useful because a scanner will allow you to listen to city, county, state, and federal law enforcement, fire departments, local bus drivers, some amateur radio frequencies (e.g., the 2M band), air traffic and marine transmissions, and more.

You can probably imagine what you would hear on a scanner during a disaster scenario. For example, an area where there is a lot of law enforcement or emergency services activity is one you will probably want to avoid. Or you might be surprised to hear that there is no radio traffic in some areas, which could be even worse, because emergency services might not be available at all. Maybe you can determine that some areas may be free of danger or congestion. Endless possibilities aside, if you're able to listen to radio traffic in your area, you'll be able to make informed decisions versus simply speculating.

One important consideration to keep in mind: if you don't know what normal traffic sounds like, then you won't know what the difference is during an emergency. As with any piece of equipment, you need to use it! If you have a scanner (and I recommend you get one as part of your emergency communications plan), you should make a point of periodically listening to it and make sure you are familiar with your territory.

Most scanners are "empty" out of the box, which means you will need to find a way to program frequencies into them. While you could simply scan from one frequency to another, covering every single frequency between the two, that is a slow and inefficient way to scan. It is much more effective to scan specific, *known* frequencies in a continuous loop, stopping when traffic is heard. And in case you're wondering, these frequencies are not secret. They are publicly available and legal to listen to.

One great resource for finding frequencies in your area is www.RadioReference.com. Taking a look here before you invest in a scanner will give you a good idea about what kinds of traffic you'll be able to listen to. Of course, if you live in an urban area, there will be plenty to hear.

One more thing to consider is that many government agencies are slowly converting their systems from analog to digital. You may want to find a scanner which can also intercept digital transmissions, though you should expect the price will be higher.

In my opinion, most scanners are a hassle to set up and use, and unless you are committed to going through the setup process, you may just throw up your hands in frustration. But keep reading — there is hope!

I found one exception (and please send me an email if you learn of any others), the **Uniden Homepatrol** (Figure 6-1). This is a truly unique scanner, and when I bought it, there was nothing similar available.

Not only does it have a touch screen with intuitive and responsive on-screen buttons (good job, Uniden user interface designers!), even if you have no scanner experience, you can easily set up as many types of frequencies as you like in just minutes.

Here is my personal experience. I opened the box, plugged in the power supply (it can also recharge AA batteries or run on disposables), and connected it to my computer with the included USB cable to synchronize with the Internet database. In just a few minutes I had selected all of the groups of frequencies I wanted, and **I was actively scanning and listening.** That is an easy setup!

My experience with another high quality scanner went like this: I unboxed the scanner and started charging it, then went online looking for a database with up-to-date frequencies. After some research, I eventually found a database that allowed access with the software I would get later. For a fee, I got access to the database and its updates for a year. Then I paid for the special software required to communicate across the special cable, after connecting to the special database. Several hours later (over the course of

FIGURE 6-1: Uniden Homepatrol, probably the most user-friendly scanner currently on the market.

a few evenings of learning curve), I had some frequencies programmed in my scanner. I was not looking forward to updating them. And this was a good scanner, not a piece of junk. As with many technologies that aren't cutting-edge, this manufacturer was still living in the Stone Age when it came to using a computer or the Internet to make the user's experience easier.

The differences are striking. If you have the budget (they are priced at around $500 at this writing), the Homepatrol is the only scanner I can recommend with good conscience to anyone other than a scanning aficionado (who might need specialized or advanced features that may be available in other models). You can find more details and pricing online.

If you are on a tighter budget or want the adventure of programming a scanner on your own, you can find various used scanners on the market. If you need help setting yours up, I suggest you either find someone in your area who can help in person (a local ham club would probably be a good starting point), or join an online user group.

If you are technically minded, a scanning enthusiast, and interested in different features or sizes of other scanners, of course you'll want to look at the many other models available. There are very powerful and usable scanners on the market. But for beginners or anyone who wants something simple (like me), it's tough to beat the Homepatrol. You can also find more information on setup, specs, etc., at www. HomePatrol.com.

Note: if you know of a scanner that is reasonably simple and intuitive to configure and use, especially one that costs less than $500, please send me a note. I hope that Uniden's innovative design sets the trend with other manufacturers.

COMMUNICATING IN TWO DIRECTIONS

I'll go out on a limb and assume that since you made it this far in the book, you'll understand that in addition to having a radio available to listen to in an emergency, being able to talk to someone would also be a good idea.

Do you have family, friends, neighbors or anyone else you really care about? Very few people are truly alone when it comes to personal relationships. Even if you're geographically distant from loved ones, you aren't far away when it comes to radio waves, which can travel tens, hundreds, or even thousands of miles at the speed of light.

Are you ready to move to the advanced stages of emergency communication preparation? I have great news: setting up two-way communications can be incredibly easy and inexpensive!

CHAPTER 7:
FRS & GMRS RADIOS

The snow was falling even more heavily. Bruce guessed that it was probably an official blizzard now. He couldn't see more than a few feet out the window. Not even seasoned locals had expected this much snow, though they were quite used to cold weather in this part of Maine. Although the power had gone out earlier that morning, his generator was running smoothly and Bruce knew everyone in the neighborhood was well prepared for bad weather.

Nonetheless, he couldn't help but wonder about Mrs. Patterson, his closest neighbor, who lived about a mile down the road to the north. Her husband had died the previous year and Bruce's family had been helping her out with yard work and other heavy tasks ever since.

Though Bruce doubted Mrs. Patterson would admit it, he knew from earlier conversation with her late husband that she was diabetic, and that her health was slowly deteriorating. Not only that, she had begun forgetting things more frequently, like where she had left her keys or her phone.

Bruce's mind wandered as he stood at the stove and slowly stirred a pot of chili. He thought back to what happened last winter. An elderly man on the other side of town froze to death after his power went out in the night and his electric heaters stopped working. With no local family or neighbors to check on him, it was several days before the police found him.

Bruce shuddered and decided to make a call. He scanned the list of numbers stuck on the bulletin board and dialed Mrs. Patterson's phone number. She was probably fine, but he wouldn't feel good until he checked.

And since the telephones often worked even when grid power was down, they could probably still communicate.[7]

There was no answer, which was unusual. She was always home in the evenings, especially when the weather was bad. He knew she had a wireless landline phone, a gift from her grandkids, she'd told him. Maybe she had set it down and let the battery run out, or forgotten where she'd left it.

Bruce looked out the window, as the dark, gray evening turned black, the snow falling harder now. He didn't feel like gearing up and driving the snowmobile a mile in a blizzard, especially without good reason. And at this point, an unanswered phone wasn't reason enough. It was time for Plan B.

"Mary, we might have a problem." His wife Mary was sitting in the living room, concentrating as she typed on her computer.

She could tell by his tone that something was wrong. She stopped typing. "What's up?"

"Mrs. Patterson isn't answering her phone and I'm a little worried about her. And the snow is coming down even harder now. I'm probably just being paranoid, but would you keep trying her on the phone while I try the radio?"

"Sure."

Bruce handed Mary the cordless phone, and she hit the redial button while he headed into the living room. Grabbing one of the FRS radios from the bookshelf, where it had been sitting quietly, he turned it to channel six, and pressed the transmit button.

"Mrs. Patterson, are you there? This is Bruce."

Earlier that summer, during a brief power outage, Bruce had thought about how he could communicate with his aging mother and father who lived a couple miles down the road to the south. Then he read a book

---

[7] Your landline phone is powered separately from your AC grid power. Telephones use DC power flowing over the phone lines themselves. This means you may be able to make a phone call even if the power is out. However, if you have a wireless handset that requires AC power to transmit from a base to a handset, it will probably not work when your power is out. Make sure you have at least one telephone (with an old-fashioned cord from the handset to the base) that can operate only using the power supply transmitted over phone line.

on emergency preparedness, which had some clear recommendations on emergency communications. He decided that an inexpensive set of FRS/GMRS radios would be a good backup plan if the phones failed.

When he went to the local sporting goods store to buy a pair of radios, he picked up a second pair on a whim. He gave one radio to his parents and kept two for himself so he'd be able to talk with his family if he needed to leave the house. He gave the other one, with its own charger, to Mrs. Patterson. He showed her how to press the button on the side and talk into the built-in microphone. It was pretty simple. He plugged it in for her, set it to channel 6, using a specific privacy code to exclude the idle chatter of anyone else in the area who might have been using the same channel, and left it on. Since she wasn't close to anyone else using this type of radio on that channel, it didn't disturb her. In fact, Bruce doubted she remembered it was there. It sat on her kitchen shelf trickle-charging, waiting for Bruce to call. She was ambivalent about it, but her having it made Bruce feel a lot better. And now, even if her power was out, it should be in standby mode, running on battery power for at least the next forty-eight hours.

He had taped a small piece of paper with the channel and privacy code for Mrs. Patterson's radio, as well as the one for his parents, to the back of his own radio. He double-checked the setting. Yes, channel six was for Mrs. Patterson. He called again.

"Mrs. Patterson, this is Bruce, your neighbor. Are you OK?" Bruce paused, listening, dreading the idea of venturing out into the snowstorm. "She's not answering the radio either," he called out to Mary. "I think I'll need to head over there." He tried Mrs. Patterson once more. There was no answer.

Bruce took the radio with him to the bedroom and continued to listen as he stripped, then pulled on long underwear, pants and shirt, and then his snowsuit. Then a hesitant voice came over the radio.

"Hello… who's there?" The voice was feeble and difficult to hear, not because the signal was weak but because she was speaking quietly. Mrs. Patterson was clearly alive, but didn't sound right.

Alarm bells rang in Bruce's head. "Mrs. Patterson, this is your neighbor Bruce. Are you OK? There's a lot of snow out there, and I wanted to make sure you were safe." Bruce's wife entered the living room and stood silently, listening.

"I want to call Bill, but I can't find my telephone."

Mary and Bruce looked at each other. Bill was Mrs. Patterson's dead husband.

"She's really confused," Mary whispered. "I've never known her to be this bad." Bruce pressed the transmit button again.

"Mrs. Patterson, are you OK?"

"I need Bill to fix the heater. It's so cold in here. And my electric blanket doesn't work. And the bed is so cold…"

Bruce's heart leaped. This was serious.

"I'm coming over, Mrs. Patterson, right now. Can you please make sure the door is unlocked? I'm going to bring you a portable heater, so you don't get too cold."

Bruce looked at Mary. "Call 911. We might need the medics out here, if they're able to make it. She's probably hypothermic, or will be by the time I get there. Then call my folks. Worst case, Dad can run their snowmobile safely in this kind of snow if I need his help for some reason. I'm going to take the little propane heater. It's probably freezing over there now."

He rushed to the garage to grab his large backpack. Their "Mr. Heater Little Buddy" propane-powered heater fit inside it easily, along with two additional propane canisters, a spare blanket, Mary's down jacket, a fleece hat and thick socks.

Minutes later, Bruce was suited up and ready to go. His pack was on his back and the snowmobile was warmed up and purring in the driveway, and he had a radio in each hand. He handed one to Mary.

"Keep monitoring channel six." He didn't mention the possibility of getting lost, but knew it could happen. The radio would give him a way to keep in contact in any case. He called again.

"Mrs. Patterson, I'm coming over. Please unlock your door." He wasn't worried about breaking one of her windows if needed, but figured it wouldn't hurt to make sure. She didn't reply. Bruce moved faster now.

He tucked the radio into an inside pocket in his snowsuit, zipped up, and kissed Mary on the cheek as she filled in his father on the phone. He strapped on his helmet, and stepped out into the whirling snow.

FRS and GMRS radios are currently very common in the US. At the time of this writing, it's estimated there have been up to 50 million units sold, maybe even more. That's good and bad, but the good definitely outweighs the bad. The good part is that you'll likely be able to reach someone (nearby, at least) in an emergency. The bad part is that the channels could be so busy as to render them unusable (depending on the population density where you live). You may not be able to hear or transmit a critical message. Nonetheless, you should have at least one pair.

Here are more pros and cons you should think about if you are factoring FRS/GMRS radios into your emergency communication plan.

Pros & Cons

Pros:

- FRS/GMRS radios are easy to find. You can find them at most outdoor and sporting goods stores, or at dozens of reputable retailers online.
- No license is required for FRS, although one is still technically required to operate GMRS.[8] This means you can buy radios, put in batteries, and get on the air right away.
- The cost is usually relatively low. Some of the least expensive, less powerful radios can be purchased for around $20 on Amazon.com. That is an inexpensive two-way communications solution! And if you pay a bit more, you can get more powerful and higher quality radios.
- You can easily find someone to talk with (whether you want to or not).
- These radios are generally flexible to use. They're handy for close-proximity activities, such as on a camping trip or in a convoy. In an

---

[8] Per the FCC (at www.FCC.gov in early 2012), "In 2010, the FCC proposed to remove the individual licensing requirement for GMRS and instead license GMRS 'by rule' (meaning that an individual license would not be required to operate a GMRS device). This proposal is still pending."

emergency, they can be repurposed for close-range communications, e.g., to communicate with neighbors.

- They usually run on AA batteries. *This is important.* During a power outage, you may not have the ability to recharge an internal battery pack. If you have a stock of AA batteries (and you should — see emergency power options in Chapter 15), you will be able to talk on the radio without a charger. *I recommend you avoid buying FRS/GMRS radios that can't operate on AA batteries.*

Cons:

- No license is required for FRS channels, which means you'll have more people who don't know how to talk on the air, noisily and inefficiently occupying these frequencies.
- Limited power: FRS radios cannot exceed 0.5 watt (500 milliwatts), which is not a lot of power. In addition, they can't use a separate (more efficient) antenna, which prevents you from improving their receptions/transmission capability. Handheld radios that are GMRS-only (without FRS built in, and which are far less common) can transmit at up to 5 watts, and use detachable antennas (per FCC guidance), however, when they're integrated in the same radio with FRS channels, you are stuck with the small, integrated antenna, which will limit their range (although they can still output up to 5 watts).
  - Note: Mobile (vehicle-mounted) or base station (e.g., on your desk) GMRS radios can transmit at up to 50 watts, and even use repeaters (depending on the model), but this is uncommon. You can investigate further on your own.
- FRS/GMRS radios are limited to 22 channels,[9] which can make them nearly useless if too many people (which will largely depend on your location) are trying to talk at once. This means your important message could go unheard.

---

[9] Some manufacturers claim to have additional channels available, but they are using the same channels along with hard-coded special tones, similar to using privacy codes. 22 channels are still all you get.

- GMRS requires an FCC license, which costs $85 for five years, according to the FCC in September of 2011 (although there is currently discussion on eliminating this requirement. And in any case, the requirement appears to be largely ignored, as with CB radio in previous decades when a license was required).
  - Note: In comparison to GMRS licensing, amateur radio licenses are free (with a small exam fee, more in Chapter 13.)
- As I mentioned earlier, range can be significantly limited by anything between you and the person you're trying to contact. If this type of radio is part of your emergency communication plan, you must test to ensure you can cover the distances you anticipate needing to cover.

Recommendations

I recommend that every family have at least one pair of FRS/GMRS handheld radios. Two sets is even better because it gives you the option to loan radios to neighbors (or other family members, depending on family size). Two significant advantages make these radios worth owning:

1. They are easy to find, buy and use. They are relatively inexpensive, easy to find in many sporting goods or consumer electronics stores, and can usually run on common, disposable batteries.
2. You can use them for intelligence gathering. You can simply listen and learn more about what is happening in your vicinity. Since their range is limited, anything you hear is happening relatively close by. Remember, we're talking about an emergency situation. Everyday non-emergency traffic usually consists of kids playing. However, in an emergency, you should expect the traffic to become much more serious.

Quality brands include Midland, Motorola, Cobra and Uniden. I suggest you get a waterproof (or at least water resistant) model that also allows you to use privacy codes. These codes allow you to filter out traffic on a certain frequency if they aren't using the same code.

Note: if a frequency or channel is overloaded with traffic, a privacy code will not make it any easier for you to hear your traffic. It may still be drowned out by the other transmissions.

The Midland radio shown in Figure 7-1 is one of a set I own. They came with an AC charger (a "wall wart" and drop-in charging base) as well as a 12-volt car charger (a critical accessory), and headsets, which are convenient for you and others in a noisy environment.

Warning: Don't be fooled by aggressive marketing (almost all the manufacturers are guilty of this practice) of FRS/GMRS radios' effective range. If they were ever actually able to reach each other from thirty-six miles away, as

FIGURE 7-1: Midland GXT 850 FRS/GMRS radio

some manufacturers boast, it had to have been with a completely unobstructed line of sight, for example, one hilltop to another with nothing in between, or across a lake. In general, when you have trees, buildings, or just about anything else preventing you from having a straight, perfectly clear line to the other radio, the range will be significantly decreased. It's common to only get half of the advertised range, and much less in urban areas. This is normal. Don't get discouraged or feel ripped off. It's simply the nature of these types of radios.

One other interesting option that exists for FRS/GMRS is the Midland XT511 hand crank radio (Figure 7-2) that also receives AM, FM, and NOAA weather alerts. To extend the feature list further, it has a flashlight and USB port for charging other devices. It has five power sources: AA batteries, a separate, rechargeable battery pack, AC and DC adapters, and the hand

FIGURE 7-2: Midland XT511 hand-crank emergency radio.

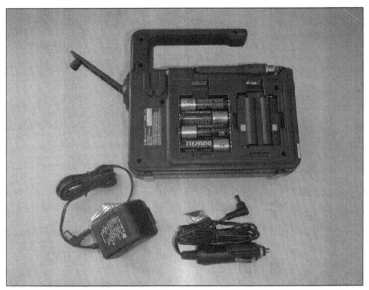

FIGURE 7-3: You can see the various power options for the XT511. It has an AC wall-wart (bottom left), auto 12-volt adapter (bottom right), four AA batteries (inside radio, left), a rechargeable battery pack (inside radio, right), and the charging crank (sticking out to the side).

crank. Having an option that will work when both AC and battery power fail is a good idea, and having AM/FM/NOAA reception reduces the need for yet another radio.

There aren't many radios with this much power supply flexibility. All of these options will give you a more options in an emergency.

I found an XT511 for $56.00 on Amazon.com. If you are just starting to figure out what you need for emergency communications, I recommend that you seriously consider the XT511.

CHAPTER 8:

CB – THE CITIZENS' BAND AND TRUCKER'S FRIEND

"What do you mean, you forgot to get gas?" Kayla demanded, gritting her teeth. "We were just at a gas station!" Her boyfriend shrugged.

"I was hungry and thirsty and had to pee," Paul replied. "So I forgot. Don't make such a big deal out of it!"

Kayla snapped her phone shut in exasperation. "No coverage. What do we do now? Wave someone down and ask to borrow a gallon of gas?" Her voice was pitched higher now, and her body language made it clear that she wanted nothing to do with Paul.

Paul glanced at his watch: 10:20 PM. Considering the late hour and the fact that his 1995 Toyota Tacoma had run dry on a deserted Forest Service road, flagging someone down didn't strike Paul as a realistic possibility. Neither was walking back to the highway, which was over ten miles away. Not on a cold February night anyway, and although the road was drivable, it would be rough travel on foot.

Paul drove out there hoping for some romantic time in the moonlight at one of the most breathtaking scenic overlooks in the area.

"We could spend the night," Paul said, weakly.

"In your dreams," Kayla replied. "And my father will murder you if I'm home late. Especially if it's because you nearly got me killed."

Paul racked his brain. They couldn't stay. They couldn't walk back. They had no gas. Neither of them had cell phone coverage.

"Does your radio work?" Paul gave Kayla a blank look. "Your radio, dummy. Does it work?" Kayla reached over and flipped the power switch on. Static erupted from the speaker.

Paul bought the truck only two weeks ago, and while he'd noticed the radio and its cool-looking antenna (one of the reasons he'd liked it), he'd never turned it on and had no idea how to use it.

"Uh, I don't know."

Kayla gave him a withering look and looked at the radio more closely. She picked up the hand microphone, pressed the transmit key, and said "Is anybody out there? We need help!"

Luckily for them, the previous owner had left the radio set on channel 9, the emergency CB channel. And, since their path had been roughly parallel to the highway, they were not far from anyone travelling on it.

"Hey there, little lady, this is Roger Dodger, on Interstate 90 headed east. What's your situation? Are you OK?"

"My name is Kayla and this idiot I'm with," she glared at Paul as she spoke, "ran out of gas on a Forest Service road, and we need some gas to get out of here so we don't freeze to death. Can you call someone to help us? Our phones don't work."

"Sure thing, Kayla. I'm pulling off the road right now so that I can stay in range. Give me a minute to make a call. I'll see if I can get the State Patrol. Hold tight."

"Will do, Roger Dodger. Thanks a bunch!" She turned to Paul with another withering look. "It's your lucky day."

Paul shuddered, not from the cold, but at the thought of what Kayla's father would have done if he hadn't brought her home safely...

Were you alive in the 1970s? If so, you probably couldn't avoid the CB excitement that swept through Hollywood and the rest of the country, with truckers fighting biker gangs, Robin Hood-esque outlaws foiling the efforts of bumbling sheriff deputies, and other such fascinating plots. That was the peak of CB radio's popularity.

Nonetheless, it's not uncommon to find truckers and certain businesses that still use CB radios. Many hunters and outdoors aficionados also sport the CB whips on their four-wheel drive vehicles. Depending on where you live, you may have a healthy user base with whom you can communicate in

FIGURE 8-1: Older CB radio base station.

an emergency. In addition, depending on your emergency communication plan, you may only need to communicate with a few people locally. If some or all of them have CB radios in their vehicles or homes, you already have a piece of your plan in place!

One thing to keep in mind, depending on your area, is that some people seem to enjoy getting on their CB radios only to curse at each other, use illegally amplified signals to drown each other out and engage in other inappropriate behavior. You are far less likely to find such behavior using amateur radios, at least in my experience. If you have a CB radio and want to simply listen in, it may not be family-friendly conversation, so be careful.

One significant advantage with CB is channel 9, the dedicated emergency channel. Some radios allow the user to constantly monitor channel 9 for traffic (whether the radio is being used on other channels or not), which means that in an emergency, if you can transmit on channel 9, you may have someone out there waiting for your call.

The alternative to a dedicated channel, whether using CB or any other type of radio, is to scan until you find a conversation, wait for a pause and say "Break, break" (the abbreviated version of "Please let me cut in and say something really important,"), and when you have a couple seconds of silence, explain your emergency.

Pros & Cons

Pros:

- Gear is relatively inexpensive
- No license is required.
- Radios and antennas are relatively simple to set up and use.
- Channel 9 is dedicated for emergency use.
- If you are near a widely used highway, there are probably truck drivers monitoring who will relay a message in an emergency.
- If you have a modern CB radio, you can transmit and listen using upper or lower sideband,[10] which effectively triples your channel count from 40 to 120, making it easier to find a free channel to use if the airwaves are busy.
- Sometimes, depending on atmospheric conditions, you can communicate across long distances (although you cannot reliably depend on this ability, and it actually may decrease your ability to communicate locally — see "Cons" below).

Cons:

- You can only use a set number of channels (as with FRS/GMRS and some other radios), although when using upper or lower sideband, you use up to 120.
- CB is not popular in some areas.
- Depending on atmospheric conditions, you may not be able to communicate locally, as your signals could be "skipping" far away (bouncing off the ionosphere), which makes CB radio less reliable than some other solutions.

---

[10] SSB, or Single Sideband is an option on better CB radios which allow you to use an upper or lower mirror image (Upper Sideband or Lower Sideband, shown as USB or LSB) of the original channel frequency. Since this method uses power more efficiently, it allows you to transmit at greater ranges, and effectively increases the CB channel count from 40 to 120. As long at the person you are speaking with also has the LSB/USB options, you'll be able to communicate with more flexibility and possibly more range.

- Handheld CB radios still require an antenna, and an effective antenna for CB frequencies will be long, making portable CB communications inconvenient in many cases, as compared to FRS/GMRS or amateur UHF/VHF.

Recommendations

Depending on your specific needs and whether you have many other users in your area, CB may be an easy and useful option. Not only are radios relatively inexpensive, but you can also frequently find used ones for sale. You may also want a CB radio just for listening to local traffic (intelligence gathering). Since it's still a favorite form of communication for many, it's a good idea to consider CB as a part of your emcomm plan.

CHAPTER 9:

EXRS — FREQUENCY - HOPPING, SPREAD SPECTRUM, HIGH -TECH COOLNESS!

Jefferson and SueEllen each took charge of two of their four kids — Jefferson took their two boys, Dustin and Cody, ages seven and nine, and SueEllen took their two girls, Bethany and Veronica, ages four and five. Conveniently, the boys were both more interested in the death-defying rides, while the girls were drawn to rides that spun in circles.

As much crazy fun as Wally-World was going to be, SueEllen took the planning very seriously. They had two check-in times, when they would meet first for an early lunch and later for an afternoon snack. This way, she reasoned, they'd cover everything the kids wanted to do, while making time to share their experiences as a family, if only briefly, and it would ensure they wouldn't be separated for too long in such a chaotic environment. Since the kids didn't have cell phones yet, and nobody was looking forward to the extra expense on the already overloaded phone bill (two pages of taxes and fees!), they were holding off for as long as possible. Instead, everyone had a handheld, two-way radio. There would be no monthly bill for the radios, no matter how often they used them!

"Does everyone have their radios?" SueEllen asked. "Check in, please."

"Dustin here."

"Hi, this is Bethany." Cody and Veronica also checked in, clearly audible.

As they made their way into the crowded park, it appeared that many kids and grown-ups had radios in their hands or clipped to their belts, and there was a lot of noise. Everyone, it seemed, was using a FRS/GMRS radio, which offered a limited set of channels to share. That meant that many people were using those few channels at the same time.

SueEllen and the girls parted from Jefferson and the boys, and each group disappeared into the crowds, eager to make their way to the rides.

An hour later, Jefferson waited for Dustin and Cody to get off the Monster-Coaster. The car had just come to a stop, and people were filing out. Even from several yards away, he could see the ear-to-ear grins on their faces. Jefferson grinned too, glad he was able to give his kids this amazing experience.

His warm thoughts were interrupted by a shout, one word that stopped him in his tracks, like a deer in headlights.

"Fire!"

Jefferson spun around, looking for signs of smoke or flames, but saw nothing. He did spy a Wally-World employee with a fire extinguisher running toward the spinning teacup ride. But he saw something else that shocked him, a tidal wave of people moving swiftly toward him. Before he could react, he was swept along, away from the roller coaster and his boys. The cries of "Fire! Fire!" continued, and panic ensued.

Like a salmon struggling to swim upstream, Jefferson fought his way through the crowd. Finally, he was able to duck behind a pillar and wait for the rest of the crowd to push by. He heard screams, and saw several people lying on the ground, some motionless and others clutching injured limbs, crying. One woman who was trying to get to her feet had blood streaming from her nose.

Cody and Dustin were nowhere to be seen. There was still no smoke visible and for all he knew, the initial blaze had been put out right after it started, but the chaos was still spreading through the rest of the theme park.

Jefferson keyed his radio and called, "SueEllen, take cover! Get the girls and get out of the open. There's a panicked crowd heading your way!"

"I hear you, Jeff. I'm doing it now!"

"Cody! Dustin! Where are you?" Jefferson shouted as he sprinted back towards the roller coaster. There was no sign of them. He tried his radio. "Cody, Dustin, where are you? Are you OK?" He changed direction, jogging back the way he'd come, wondering whether Cody and Dustin had been caught up in the panicked flow.

"We're here, Dad!" Cody's voice crackled across the radio. "I'm with Dustin, and we're OK."

"Where are you?"

"We're by the roller coaster."

"Wait there for me. SueEllen, did you hear that? Are you OK?"

"Yes," SueEllen replied. "We're in the gift shop, but there are still people outside freaking out and running around. I think most of the crowd has already passed, and it looks like a few people are hurt, too."

"Just wait there. I'm going to find the boys and we'll come meet you. You're not far from the exit."

As he quickly made his way toward the roller coaster, he passed a man yelling into a radio.

"Janie, I can't hear you! Are you OK?" When the man paused to listen, Jefferson noticed, all that came across was the clamor of a dozen other parents and kids all talking at the same time, voices filled with fear. "Janie, are you OK?" he repeated, shouting into the radio he clutched, white-knuckled, with both hands.

Jefferson pressed the "Transmit" button on his own radio. "Is everyone still OK? Everyone check in with me now, please."

"This is Cody. We're still OK, but a security guy came by and said we had to go to the exit. We're walking with him toward the green gate."

"This is SueEllen. We're OK, too, and it looks safe from what I can see. I'll take the girls toward the green gate by the exit, too."

"OK, I'm headed that way now," Jefferson replied. "Let me know right away if anything changes."

The scene near the gate was only a little less chaotic than it had been earlier. The flashing lights of emergency fire and medical crews added to the overall confusion, and Jefferson saw more people shouting into radios, unable to reach loved ones.

Four minutes and two more check-ins later, the family was united, unhurt but shaken.

"I'm glad you bought these radios, honey," Jeff said. "They worked when all those other ones didn't."

Do you have anything private to discuss on the radio? Are you concerned about having anyone being able to listen to your radio conversations? Are you tired of hearing neighbors, vacationers, and random children constantly blabbering on all the channels of your FRS/GMRS radio when you have urgent business to discuss?

If so, you may be glad to hear that high-tech, frequency-hopping, handheld radio communications are available without a license. And while FRS/GMRS radios usually have only 22 channels, a relatively new radio system has ten billion! To learn more about this amazing, little-known technology, keep reading and prepare to be impressed.

The Problem

Every communication option has its pros and cons. The easiest and least expensive option is usually the FRS/GMRS radio, and that's what many people already have. But for roughly the same price and effort (namely, taking the radios out of their packaging, charging them, setting a channel and talking, with no license needed) you have another option which changes the game significantly.

Let's quickly review one key limitation of the popular FRS/GMRS option. Since this system has a limited number of channels and these radios are so common, in an emergency you should expect to have noisy competition for every channel. In addition, you may want an option otherwise unavailable in *any* other system, whether FRS, CB, marine, aviation, amateur radio, or even public service (police,[11] fire) frequencies (which you can listen to on a scanner). What is that option? Privacy.

None of the other options mentioned allow you to speak with practical privacy. Anyone with a similar radio or a scanner can easily listen to your conversation.

---

[11] Due to the continuously decreasing cost and increasing protocol standardization, more police agencies are using encryption on certain channels, making legal interception/scanning impossible. However, at the time of writing, most police radio traffic can still be easily intercepted.

Of course, you're not going to talk about moving your million-dollar gold hoard from your vault to a friend's house two blocks away, but what if you needed to discuss sharing emergency supplies with a neighbor, and you live in an area with gang activity? We all wish that every neighbor was friendly, but most of us have neighbors with whom we don't get along. And in a desperate situation, one of your neighbors may not think twice about liberating you of anything they want or need. But if you find any of those scenarios unlikely, you still may prefer privacy for privacy's sake.

If you only have FRS or amateur radio, assuming your channel or frequency is available in the first place, you can hide the meaning of your conversation by talking in code. Per FCC rules, this is not allowed on amateur radio frequencies, (although it is unlikely the FCC will be enforcing such rules during the time of a disaster). You may even attract unwanted attention with this approach, unless your code-talk sounds like normal conversation. This also means you and whomever you are talking with will need to remember all of the code words or phrases, which will be much more difficult if you are stressed out! Talking with your everyday vocabulary will be much easier. There must be an easier way. And there is…

The Solution

We have a solution: a technology called "Frequency-Hopping Spread Spectrum." It can't be effectively intercepted, although I have heard the FCC does possess the technology to intercept this type of communication, using special equipment not available to the public.

The traffic is not actually encrypted. Instead, it's simply transmitted in a different, unusual way. Using a technology developed in the early 1900s, these radios constantly, simultaneously switch from one frequency to another during the transmission. So if you were trying to listen in by monitoring any single frequency, which is what a typical scanner will do (one frequency at a time), all you would ever theoretically hear is an occasional, infrequent, microsecond blip of noise. However, if you had a compatible radio and the code to synchronize them, you would hear the

FIGURE 9-1: TriSquare eXRS TSX300, an inexpensive, flexible and private communication solution.

whole conversation! But nobody else could hear it all.

I am aware of only one brand of radio readily available to the public that uses this technology. It is the TriSquare eXRS system, and if you need to be able to talk without interference, don't want casual listeners to intercept your conversation, and want a lot of other interesting options built in, you should consider getting a pair of these radios.

Additional Specs

These radios use the 900 MHz frequency band, which will limit the user's range in some ways, similar to how FRS/GMRS radios are limited. Different radio frequencies will have different **propagation** characteristics, meaning the radio waves can travel farther in certain cases, or are limited by different materials, such as forests, crowds or walls. These radios put out one watt of power, which is relatively low (although twice as powerful as a 500-milliwatt FRS radio, and probably less powerful than a GMRS handheld radio, which is usually between one and five watts).

To be clear, as with FRS/GMRS radios, these are not long-range, powerful ham radios. These are short-range radios, useful for a conversation within your neighborhood, between nearby buildings, cars in a convoy, etc. Don't expect them to work twenty miles away (and probably not even ten miles away), unless you have line of sight conditions, meaning that if your eyesight were good enough, you could actually see the other radio you want to speak to, with nothing blocking the view.

In my experience, eXRS radios work for a couple miles in the countryside, and for two to three hundred meters in high-rise, steel and concrete city territory, depending on what's in the way. Of course, if you don't have buildings between radios, the range grows considerably. Nonetheless, they still have some incredible features and are definitely worth considering, depending on your circumstances. More interesting details follow.

From the TriSquare website (www.TriSquare.us):

"[The eXRS radios have the] added benefit of wide band digital security and privacy provided by the frequency hopping spread spectrum algorithm. Maintaining the information signal as narrow band FM modulation centered on discrete frequencies allows for a large pool of non-overlapping hopping frequencies to draw upon within a given section of the radio frequency (RF) spectrum. The pseudo-random drawing of the hopping frequencies spreads the total signal power equally over the entire bandwidth of the RF spectrum used, which ensures minimal interference between many simultaneous, independent users.

Interference Free is a major advantage eXRS has over existing FRS and GMRS radios. Based on a contracted study from the Electrical and Computer Engineering Department of a major university, *more than 100,000 eXRS users within talk range can enjoy uninterrupted communications. Whereas, FRS/GMRS quickly becomes unusable with just a few tens of users within range of each other. Bottom-line is that the eXRS two-way radios provide private communications in areas where FRS/GMRS conversations can be congested or impossible."* [Italics mine.]

Let's talk features. The radios are not set up to talk on FRS/GMRS frequencies, and can only communicate with other eXRS radios. However, they can receive NOAA weather radio transmissions, which is important (see earlier discussion of why you need NOAA in Chapter 5).

Here are other available features, some of which I have not seen on any other radios:

- Select a group or an individual, depending on who you want to talk with.
- Identify any radio with a ten-digit number (which gives you about ten billion options!)
- Identify any specific contact or group of contacts using unique alphanumeric codes.
- Page/call a specific radio or person using a unique tone (five tones are available).

- Call waiting.
- Voice-operated (VOX) operation.
- Text messaging (send whatever message you like, or use pre-defined messages). Your radio can store messages as they come in, and you can retrieve them at your leisure.
- Clone radios or transfer contacts with no cables needed.
- In addition to the rechargeable battery pack, they can run on AA batteries.

These radios are impressive.

There are two models, one with just a few keys (TSX100), and one with a full keypad (TSX300). I recommend the full keypad, because it gives you more flexibility. It costs a bit more, but I think the added ability to easily enter numbers, etc., makes it worth it.

If you are not inclined to get an amateur radio license (although I highly recommend it, since it's easy to get and extremely useful! More on that in Chapter 13), this can be a useful way to communicate in a densely populated or otherwise noisy area, or if you need privacy and text messaging features. In addition, these radios are just plain cool!

Pros & Cons

Pros:

- Privacy!
- You can communicate even if many other radios are transmitting at the same time your area.
- They are portable.
- They can run on AA batteries in addition to rechargeable battery pack
- eXRS radios are relatively inexpensive ($80-$100 for the TSX300; less for the TSX100).

Cons:

- They are short-range only, as with FRS radios (and probably less range than powerful, 5-watt GMRS radios).

- eXRS radios only work with other eXRS radios.
- They are slightly more complicated to use than FRS/GMRS radios.

Recommendations

Think about your privacy needs. If you want to have a private conversation on the radio, get a pair of eXRS radios. Also, if you are in a population-dense area, you also might want to get a set of eXRS radios to ensure you'll be able to communicate whenever you need to.

CHAPTER 10:
PERSONAL LOCATOR BEACONS

[Based on a true news story.]

Twenty-year-old injured mountain climber Jennifer Dobbs is safe Sunday morning after a twelve-hour rescue finally concluded in Rocky Mountain National Park.

Park spokesperson Nathan Farnsworth said more than fifteen people worked nonstop in their efforts to rescue Jennifer Dobbs from the bottom of Lansky Gorge.

Dobbs and the rest of her party of four climbers started at Black Creek Trailhead at 5:30 a.m. Sunday and planned to summit Mount Peacock later that day. However, at around 9:20 a.m., as they were ascending a steep, snow-covered slope, she slipped and fell over three hundred feet, to the bottom of the nearby gorge, suffering multiple injuries along the way, Farnsworth said.

Dobbs was fortunate to be travelling with an experienced team, and one of them, Paul Armbruster, activated his SPOT GPS Messenger device, which notified emergency services personnel immediately.

Rescuers reached Dobbs by helicopter around 10:30 a.m. Sunday. She was conscious during the rescue and in a lot of pain, she said, in a later interview. "Thank God they were able to get there so quickly. I'm not sure how long I could have lasted, stuck there in the cold, unable to move."

Dobbs was transported to Lakeview Medical Center, where she is expected to fully recover.

Personal Locator Beacons (PLBs) are useful for one-way communication, in the event you only need to make a notification. Usually a beacon is used to let someone know you're in distress, although some other devices can also send routine location updates or other short messages.

If you go skiing in a location where there's even a remote chance of an avalanche, or if you like to go off of the marked trail, you should have a PLB.

Modern PLBs transmit on 406 MHz, which is received by special Search and Rescue (SAR) satellites, with no subscription fees required. They also transmit a homing signal on a separate frequency of 121.5 MHz. While the first transmission on 406 MHz is a broad alert, the second transmission is a beacon, which allows rescuers to quickly locate someone in distress.

In addition, many modern PLBs also include GPS technology and will transmit accurate GPS coordinates, encoded within the distress signal.

If you shop for a PLB, be very careful if you find something used or of unknown manufacture date. Earlier PLBs only transmitted on 121.5 MHz, not 406 MHz, and as of 2009, those devices are no longer being monitored by satellite systems the way modern ones are.

In addition to reading the directions, you'll need to register a PLB device here: beaconregistration.noaa.gov/rgdb/. And for more information, including a count of people rescued per year since 2001, take a look here: www.sarsat.noaa.gov.

Message Transmitters

There are other devices on the market that do more than simply send a distress beacon. For example, some can send non-emergency text messages as well as a specific emergency message.

With SPOT Personal Tracker (Figure 10-1) you can easily send an SOS message along with your GPS coordinates. Depending on the situation, whom you choose to alert, and your location, this could be a valuable emergency communication asset.

The downside is that it can't *receive* any message. To determine what would happen next, you would need to do detailed, advanced planning versus simply hoping for the best after hitting your transmit button.

At the time of writing, SPOT service plans start at $100/year (with another $50/year to enable sending custom text messages), and a new device costs $100–$200.

Another device, new to the market in late 2011, is the DeLorme inReach Two-Way Satellite Communicator. Using the Iridium satellite system (see Chapter 11) and paired with a DeLorme GPS (Earthmate PN-60w), it can send and receive text messages from nearly anywhere in the world. You can see how this could be very useful functionality in an emergency.

FIGURE 10-1: The SPOT Personal Tracker or other similar devices can be useful if you need to send routine or emergency messages.

Pros & Cons

Pros:

- PLBs transmitting on 406 MHz are monitored with no subscription service.
- PLBs are designed to ensure an SOS message gets out, no matter where you are.
- The SPOT Personal Tracker device is very simple to use.

Cons:

- Most PLBs are one-way communication devices. You won't know whether anyone got your message or whether help is coming.
- Most PLBs are expensive, from $200-$500.
- The DeLorme inReach and SPOT devices are similarly priced, and come with additional monthly subscription fees. You can expect that any new technology using one-way or two-way text messaging communication services will come with subscription fees too.

CHAPTER 11:
SATELLITE PHONES

Mabel and Jurgen Schnabel had been dreaming of this trip for years and as they stood at the railing, bundled up in hats and parkas, their eyes gleamed as they took in the beauty of the icy, Alaskan landscape.

Instead of a more traditional cruise, they had opted for an unusual, smaller ship, which could carry no more than about fifty passengers, along with a small amount of cargo, and was destined for some of the more remote Alaskans, whose town piers couldn't accommodate the larger cargo or cruise vessels.

The added cost was worth it, they agreed, and even though there was no climbing wall, swimming pool, water slide, wireless Internet, or tennis court, there was a cozy, well-stocked library, spectacular views, and a much smaller, well-mannered crowd at mealtimes.

One of the disadvantages to taking the "scenic route" was they were seldom in cell phone range, and their children didn't like the idea of their parents being out in this potentially dangerous territory, out of contact for over two weeks. To make everyone feel better, knowing cell phone coverage was going to be completely unavailable after the first day, Jurgen decided to purchase a satellite phone and a small package of minutes so he could check in regularly. He settled on the Inmarsat IsatPhone Pro, and so far he had been very happy with its performance. Sometimes he'd have to move to the other side of the ship in order to get a clearer shot toward the equator,[12] but it still worked fine.

---

[12] Inmarsat's geosynchronous (technically *geostationary*) satellites hover 23,000 miles above the Earth at the Equator, halfway between the North and South Poles. To use these satellites, the user must be able to direct the phone's antenna toward the equator.

He and Mabel checked in with their son or one of their daughters briefly every day or two, and the reception was so good that they sounded like they were close by, instead of in Texas. Jurgen pulled out the phone.

"It's as good a time as any to check in." He pressed the speed-dial button for his son in Dallas. Just as he said hello, Jurgen couldn't help but notice the penetrating gaze of one of the ship's crew who had been standing nearby. The crewman strode toward them. Something didn't look right.

"Hey, Son, I have to go. We're all OK, and I'll talk to you soon. OK, I love you too. Bye."

Jurgen lowered the phone as the crewman stopped in front of them, nervous tension clear in his body language.

"Can we help you?" Mabel asked.

"Ma'am, Sir, I am sorry to bother you, but the captain would like to speak with you. Would you please come with me now?"

Mabel and Jurgen glanced at each other apprehensively before following him to the bridge, where the captain sat in a swivel chair, a worried look on his face. He stood and shook their hands.

"Mr. and Mrs. Schnabel, I'm Captain Peterson. I have a sensitive issue I would like to discuss with you." Mabel and Jurgen exchanged looks again before Jurgen looked at the captain.

"By all means," Jurgen replied. "What's going on?"

"We've had a bit of, well, an accident in our radio room, and with the electrical system on the bridge. Don't worry, the vessel is still completely safe due to multiple backup systems. However, it seems our primary and backup communications systems have been damaged and are no longer functioning. While we are still clearly able to determine our location and receive broadcasts, we are unable to communicate outward..."

"That doesn't sound good. What if a passenger gets sick, or if we need some kind of emergency assistance?"

"Yes, it is not ideal," the captain replied. "This is why I've changed our course to the nearest town with a pier, and will be making repairs and eventually obtaining replacement communications equipment. In the meantime, I am responsible for the safety of the ship, its passengers and

> crew. And since we have no other means of communicating, and since nobody's cell phones work up here..." He paused awkwardly.
>
> Jurgen smiled. "You want to borrow my phone? Sure. Here you go. The charger is in our cabin."
>
> "I will send the first mate with you and he'll bring it back here." He shook their hands again. "You are a godsend, Mr. and Mrs. Schnabel. Thank you for your assistance."

Satellite phones are becoming more common and fortunately a little less expensive than in years past. One main reason you may want one of these high-tech devices is that most natural disasters won't affect a satellite hovering many miles above the Earth. A solar flare, anti-satellite attack, or EMP burst could theoretically render them instantly useless, but as interesting as some of those scenarios might be, they are unlikely when compared to a hurricane, flood, earthquake, or other natural disaster right around us. These types of emergencies, on the ground or within our atmosphere, happen far more often.

I found three major companies that provide satellite phones and service: Iridium, Globalstar, and Inmarsat. Because technology, cost per minute, promotions, and new satellite launches change the landscape regularly, you should do your own research before making a decision.

As you can see in the picture (Figure 11-1), the Iridium satphone is not very large, especially if you were thinking you needed to set up a portable satellite dish to use one, like in the old days! Other manufacturers' consumer model phones are usually about the same size.

FIGURE 11-1: This Iridium satphone is more compact than earlier models.

Pros & Cons

Pros:

- You can make a phone call from almost anywhere in the world.[13]
- Most service providers also provide data service, which means you may be able to get a (slow) Internet connection from your remote mountain enclave or your yacht in the South Pacific.
- Most service providers also provide SMS/text or email services.
- The phones are far smaller than in the past, and often shrink in size with each new generation.

Cons:

- Phones are expensive, generally ranging from $500 to over $1000 for a new phone, although used ones may be available for several hundred dollars.
- The calling plans are expensive, and can cost from one to several dollars per minute.
- Since the phone (more specifically, its antenna) needs to "see" a satellite, it will probably not work indoors or under the cover of trees.
- Some companies, e.g., Globalstar, are not finished deploying their satellites,[14] and as a result don't have consistent coverage. This means you may only have small windows of time in any given hour to make a phone call, and unless you go online and print out their latest schedule, which changes constantly, you will not know if it'll work until you try. This may be very inconvenient for some users, especially in an emergency. (Note: From my research, it appears Inmarsat does not suffer from this problem because its satellites are in a higher, geosynchronous orbit.)

---

[13] Some satellite phone companies do not have coverage at the North and South Poles. As most of us don't go there, this should not be a common concern.

[14] At time of writing, more launches are scheduled. Go to www.globalstar.com for updates.

Recommendations

Depending on where you live, your Internet connection needs, your likelihood of experiencing a particular type of disaster, and your budget, a satphone may be a reasonable option and you should at least consider one as part of your emergency communications plan. Depending on your carrier, you could realistically get a new phone with chargers, a small chunk of minutes (per year or per month) for under $1,000. Depending on your situation, this may be a small price to pay for peace of mind.

The ability to make a phone call when the local communications infrastructure is in ruins could be an incredibly valuable asset. Note: don't forget to keep emergency contact phone numbers in your possession (especially ones out of your immediate area, whom you will be more likely able to reach in a disaster), both programmed into the phone and printed out on paper.

If you are excited about the idea of getting a satphone, make sure you investigate your options thoroughly. Ensure that you will be able to make calls when you expect to, that you will get data (Internet, email or text) access if you need it and that you won't be paying unnecessarily high monthly or yearly service fees for something you will not need to use except in an emergency.

AMATEUR RADIO – THE KING OF EMERGENCY COMMUNICATIONS

Earlier, you read about options for two-way, emergency communication. Some of them may fit your needs, budget and abilities. As a result, you may be under the impression that you're ready to communicate if your phones go down. But there's more! The best option of all is yet to come.

What is the solution most often used by city, county and federal government emergency communications teams, as well as private relief organizations around the country and throughout the world?

Some emcomm groups have satellite phones. However, not all do, usually because they're so expensive. Some groups use FRS/GMRS radios because they're inexpensive and simple to operate. But as you now know, FRS/GMRS radios have some serious drawbacks.

What technology allows us to communicate locally, regionally, and at very long distances without requiring the use of the Internet, phone lines, or satellites? Not only has it been around longer than the other technology we've reviewed so far, but it also has a dedicated user group throughout the world, and is a hobby enjoyed by young and old.

In the next three chapters you will learn about the king of emergency communications: **Amateur Radio**, also commonly called "ham radio."

CHAPTER 12:

AMATEUR RADIO – UHF AND VHF

The Redmond Amateur Radio Emergency Services (ARES) team was in full swing, speaking efficiently and calmly, despite the tension you could cut with a knife. Janeen was busy operating a pair of radios in the Emergency Operations Center (EOC) on the top floor of the police station in the middle of downtown Redmond.

The earthquake that rattled the region three hours prior caused several dozen serious injuries as well as significant property damage. People in the EOC were responsible for coordinating efforts between various fire departments, amateur radio operators working on the ARES team, and other government and private emergency relief agencies. And since Janeen had already ensured her family was safe, her current mission was to take care of others in the city who still needed help.

"KX7AB9, this is KE7RWJ, I copy your request for bottled water at the gym, and that you have approximately one half- day supply remaining. KE7RWJ clear."

"This is KH7BXP, and I need help!"

This last call was unexpected. Usually this frequency was only used by ARES team members, but in this case someone appeared to be making a personal request.

"KH7BXP, this is KE7RWJ," Janeen replied. "Please describe the problem. What kind of help do you need?"

"I know this is an emergency frequency, but this is an emergency. My phone doesn't work, and my neighbor is bleeding. I tried to stop the bleeding and put on a tourniquet, but he's in and out of consciousness and It's a life or death situation!"

> *"I heard you have a victim bleeding, a tourniquet applied, and you think his life is still at risk. Is that correct?"*
>
> *"Yes, that's right. When can we get help? Please hurry!"*
>
> *"Where are you located?"*

Amateur radio is the undisputed king of emergency communications. Why?

1. It is easy to run and frequently tested on backup power.
2. For local or medium-distance communications, it requires no external, supporting infrastructure as with trunking radio systems or repeaters.[15]
3. Licensing is easy and essentially free.
4. The equipment is flexible and powerful.
5. It can be used for local, regional, and long distance communication.[16]
6. After radios are purchased, there are no additional costs (as opposed to a cell phone or satellite phone, for example) other than occasional repairs. This is an attractive benefit to many cash-strapped emergency services organizations, as well as everyday consumers.

FIGURE 12-1: Kenwood TH-D72A, a feature-packed, powerful addition to your UHF/VHF emcomm toolkit.

---

[15] Many systems used by law enforcement and other emergency service agencies are "trunked" systems that require a central controller to manage which frequencies are used at any time. If the central controller radio/computer fails, the entire system and all its radios will not function or will be severely compromised.

[16] To achieve greater distances with UHF & VHF frequencies, operators commonly use directional antennas, and sometimes take a more exotic approach, such as using satellites as repeaters, bouncing signals off meteor showers, or taking advantage of other atmospheric phenomena. In general, UHF and VHF radios are used for local communications, whereas HF radios are used for regional or long-distance communications.

Amateur radio is an incredibly flexible and powerful communication option if you are willing to invest a small amount of time in getting a license and learning how to use your radio. With this increased flexibility and power comes an increased learning curve, but with a little effort you can dramatically improve your emergency communication capabilities!

The most common amateur radio emergency communication options, which we will discuss here, fall into these two categories:

1. UHF – Ultra-High Frequency (known as 440 MHz or 70-cm, which refers to the wavelength) amateur radios operate in a range from 420–450 MHz.

2. VHF – Very High Frequency (known as two-meter) amateur radios operate in a range from 144–147.99 MHz.

While there are other frequencies falling into the UHF and VHF ranges (e.g., the 220 MHz/1.25-meter or the 50 MHz/6-meter bands), we will focus on these two most common ones. As you'll see shortly, ham radio provides many useful options and frequencies.

Note: Do you wonder why manufacturers don't make a radio able to transmit on ham frequencies as well as FRS and GMRS frequencies? It would certainly be convenient – one less piece of equipment to carry. But the FCC generally doesn't allow it.[17]. You also can't use CB frequencies on either ham or FRS/GMRS radios.

Though handheld ham radios are nowhere near as common as FRS/GMRS, larger ham radios set up in vehicles and in offices ("ham shacks") are common, and vastly more powerful and flexible. Let's examine some basic advantages and disadvantages.

---

[17] The FCC will only certify certain types of radios to transmit on certain frequencies or bands (frequency ranges). The certification process used to be called "type acceptance." This is part of how the FCC administers the electromagnetic spectrum. This is usually a good thing, ensuring that your garage door opener does not cause interference with a neighbor's TV, your FRS radio does not interfere with a pilot trying to land a 747, your car's key fob does not zap your grandpa's pacemaker, etc. Making sure various radios get along with each other and other devices that use nearby frequencies is important for everyone. Note: you can still find radios (or scanners) that can receive across many different frequency bands.

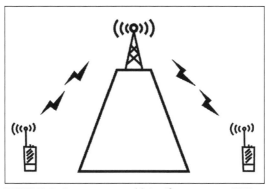

FIGURE 12-2: Repeaters can add significant range to VHF and UHF transmissions.

Repeaters – Expand Your Range

In addition to using radios to talk directly with each other, called "simplex" operation, they can be used with repeaters, called "duplex" operation. How do repeaters work? The quick explanation is that a person, ham radio club, or emcomm group installs a large, high-quality antenna in an elevated location, often on a building or tower at the top of a hill or mountain. Connected to that antenna are a radio and amplifier, which receive and then immediately rebroadcast a message on a nearby frequency. When you combine the elevation, the high-quality antenna and the high power of the amplifier, as you may expect, the signal range is far greater.

As great as that sounds, you may not be able to depend on a repeater for more than a few hours or days if grid power is cut off. It will depend on the repeater's backup power system, if it has one. You can find more explanation on repeaters and how to configure your radio to use them at www.EmergencyCommunicationsBlog.com.

Additional Details

There are other "modes" to consider with amateur radio. One interesting capability is worth mentioning for emergency communication purposes, although it's less practical because it isn't used as often. With a small,

FIGURE 12-3: The Yaesu VX-8R (and newer variations) is a very small package. It's a powerful, flexible radio and a favorite of many hams. (The antenna has been removed here for comparison purposes.)

handheld radio and a special antenna, you can receive from and transmit to certain satellites, which can then rebroadcast your signals. They act as flying repeaters. Another more commonly used protocol, called APRS, allows you to transmit GPS coordinates and other digital traffic, including text messages. The list of options goes on. Amateur radio is an incredibly powerful tool in your emergency communications toolbox.

Common amateur radio brands include Icom, Kenwood, Yaesu, Alinco, and more recently from China on the inexpensive end of the spectrum, Wouxun. The first three I listed are generally regarded as the highest quality for handheld radios. You should select a waterproof radio capable of putting out 5 watts on 2M (it will probably put out slightly less on 440 MHz). You should also get a 12-volt vehicle battery charger, a backup battery pack (most modern battery packs are lithium ion), and an AA (or AAA) battery holder. With one of these battery holders, you can insert two or three disposable AA batteries into a casing, which fits in the radio where the rechargeable battery would go. Note: you may be able to use rechargeable AA batteries in some cases, but only if they provide sufficient voltage. Many do not provide the full 1.5V per battery that disposables do. You will need to check the directions that came with your device to ensure rechargeable batteries will function.

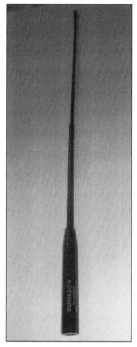

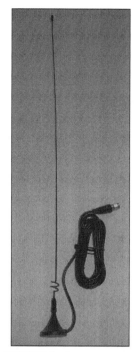

FIGURE 12-4: Aftermarket antenna for a handheld radio.

FIGURE 12-5: A mag-mount antenna with coaxial cable.

Running on AA batteries, your radio's output power may be lower, but it will work in a pinch! Everyone should have this flexibility.

Since we're talking about radios with a lot of flexibility, when it comes to using different types of antenna, I also recommend you buy an aftermarket whip antenna (Figure 12-4), which you attach to the radio directly, in place of the factory antenna. Note: almost every factory-supplied handheld radio antenna, often referred to as a "rubber duck," is quite inefficient. An aftermarket antenna can make a world of difference by providing a more efficient path for your signal.

I also recommend a "mag-mount" (magnetically mounted, see Figure 12-5) antenna to attach to the top of your vehicle (or your metal filing cabinet, if operating indoors). You attach the flexible coaxial ("coax") cable from this antenna to your radio. Such a setup will usually significantly improve your reception and transmission effectiveness.

I have described only a few things you should consider regarding ham radio, but acting on these recommendations will put you light years ahead of the person who only has an FRS/GMRS or CB radio.

We have two other options to consider with UHF & VHF radios: mobile and base station operations. Since they are usually powered by an external power supply (either from a battery or an AC to DC converter), radios designed to be mounted in your vehicle or to sit on your desk will usually have much higher power output. In addition, an antenna mounted to the top of your vehicle or high on your chimney will usually have far better reception than even a high-quality aftermarket antenna on your handheld radio. One of the disadvantages of these two options is you cannot take them with you. Some mobile units are designed to use in a backpack, but in general, they consume far more power than a small battery can provide, in addition to being larger and less portable. For portability, a handheld radio is the way to go.

If you set up a mobile or base station radio, do not neglect the antenna. Find the highest quality, most efficient antenna for the frequencies you need, mount it as high as you can, and use the best quality[18] coax possible. You will typically be able to get far better range from a great antenna than you will by pumping more power into your radio.

Most mobile UHF & VHF radios are designed to use only frequencies in the 2M and 440/70-cm bands. However, many desktop radios also transmit and receive on HF frequencies, which we will discuss in Chapter 13.

---

[18] With coaxial (coax) cable, one of the quality measures (in addition to flexibility, ability to withstand UV radiation, etc.) is its level of signal loss across certain distances. Depending on the frequencies you use and the distance between your radio and the antenna, you may want a thicker, heavier, more expensive coax versus cheap, lightweight coax. The wrong cable can cause a significant decrease in signal strength by the time the signal reaches the antenna.

Pros & Cons

Pros:

- Ham radios have *thousands* of frequency options, as opposed to being limited to a few frequencies/channels on FRS/GMRS.
- You can operate at up to 1500 watts, depending on the band (versus only 0.5 watt on FRS). This can seriously increase your transmission range.
- You can attach any size or type of external antenna to your vehicle, home (you may need a permit for a tower), or anything else, which can dramatically improve your reception and transmission ability.
- Other people operating on these frequencies:
  - usually understand how to communicate effectively with a radio and…
  - may be involved in emergency communications efforts with local or state government, the Red Cross, or others. These people are often "connected" when it comes to radio communications and may be able to help in ways that you will not find with much of the traffic on FRS/GMRS frequencies.
- Getting involved with amateur radio exposes you to emergency communication learning and volunteer opportunities, which will also provide invaluable experience (more on that in Chapter 16).
- There are many other benefits to amateur radio, which are worth exploring at www.ARRL.com, your local ham radio club, or one of the many other resources listed at the end of this book.

Cons:

- You need a license. Although it is free, you'll have to pass a basic test in order to get the Technician license, which will allow you to operate on UHF and VHF, as well as some other amateur radio frequencies. (The test isn't difficult, and you can read more about how easy it is to pass and get a license in Chapter 14.)
- Since they are more powerful and flexible, amateur radios are unsurprisingly a bit more challenging to operate, and if you want to

be able to operate proficiently, it will require some practice. Note the advantage listed above: if you are licensed, you will have the opportunity to volunteer (which means practice!) with emergency communications teams in your area.

- Ham radio gear is typically more expensive than your typical FRS/GMRS gear, although you can currently find a new, decent quality, handheld ham radio for around $100, which is about the same price as a pair of high-end FRS/GMRS radios.

Recommendations

Getting experience with amateur radio equipment and its capabilities is one of the most important ways to invest in your emcomm plan. Go get licensed, get some inexpensive equipment to learn on, and start using it!

1. Get an FCC Technician license. It is easy to study for (see Chapter 14) and the license is free. You'll get a sparkly new call sign and can join the ranks of the emcomm pros! Once you have your "license to learn," you can do and learn even more.

2. Find an inexpensive new or used, handheld, dual-band (both 440 MHz and 2M bands – this is common) radio or "HT"[19] and learn how to use it both on simplex (radio to radio) and duplex (radio to repeater to radio).

3. Get involved with *something*, whether it is a local ham club, an emergency communications group, etc. Go online and look up "ARES," "RACES," "ACS," "ACES," or "Office of Emergency Management" for your city, county, or state. You will find someone who can give you more information. If you still can't find anything, check with your local ham club and ask around. One of the mem-

---

[19] HT is an abbreviation for "Handy-Talkie." Motorola coined the term during World War II, when it introduced a handheld AM radio, the SCR-536. It later trademarked the term in 1951. Other manufacturers sometimes used the abbreviation to mean "Handheld Transceiver" in order to avoid trademark infringement.

bers will know. Another option is to contact your local ARRL section leader here: http://www.arrl.org/sections.

4. Practice, practice, practice with your radio (as with any of your important equipment). If you can't use it effectively, you're setting yourself up for trouble. Would you go buy a motorcycle you don't know how to ride, or keep a gun you don't know how to shoot on your nightstand? While there is some training available on some ways to use a ham radio (e.g., in an ARRL EmComm course[20]), "on the job" training is your best bet. Get out there and use it.

---

[20] The ARRL teaches an Introduction to EmComm course. You can find more information here: http://www.arrl.org/emergency-communications-training.

CHAPTER 13:

HF AND NVIS – HIGH FREQUENCY AND... WHAT?

"Hey, buddy, are you OK?"

"Yeah. It's pretty crazy around here, what with all the water. Most of my neighbors down the hill are long gone, because their houses are underwater. But I'm OK. The river didn't get close to my place."

Cecil sat at his desk, in front of his Icom IC-7000 radio, which was set to a HF frequency in the 80-meter band. He was speaking with his long-time friend, Rick, who was three hundred miles and several towns away.

"Do you need anything?" Rick asked. "I'm coming up later tonight. I'll be helping the Red Cross, but I can bring you whatever you need first."

A recent flood cut Cecil off from the rest of the town, but hadn't prevented him from reaching out. Even though the power was out and much of the local telephone system's infrastructure was submerged in flood waters, his radio antenna was still firmly attached to the two trees in the backyard, and the deep-cycle battery under his desk powered the radio with a lot of power to spare.

"I'm OK for now," Cecil replied. "I have a couple of batteries charged to run the radio, and the woodpile will get me through the next couple of months. I may get tired of chicken soup and chili, but I have a lot of it, and a big barrel of water. I should be fine."

"Good to hear," Rick said. "There are a lot of folks in your area who are really hurting. We'll be trucking in a lot of supplies, and we'll have to distribute some of them by boat, too, but it'll take a few days."

> *"Well, make sure you take care of other folks first. But if you get a chance to swing by, you can tie your boat to the stop sign halfway up the hill. It'll be dry after that point, and I'll have a pot of coffee heated up for you on the wood stove."*
>
> *"Will do, buddy. I'm really glad to hear you're OK. If I hadn't been able to talk to you, I probably would have come to your place first."*
>
> *"Good to know," Cecil chuckled. "Come by when you have time."*
>
> *"All right. Take care of yourself. When I get into town, I'll be using my handheld radio. You can reach me on 146.380 if you need anything. And I'll monitor this frequency for the next several hours, until I leave. We're just about packed."*
>
> *"OK, thanks. I'll let you get back to work. Talk to you soon. KG4AC clear."*
>
> *"Will do. Stay safe. KL9KHY clear."*

Amateur radio provides another very useful emergency communication option. High-Frequency (HF) amateur radio operates in several frequency bands ranging from the 160M band to the 10M band. This range is approximately 1.8 to 29.7 MHz (the low end of the 160M band to the upper end of the 10M band).

Why does HF matter for emergency communications? Because some of the HF bands, namely 40M (forty-meter), 80M, and 160M bands can be used with an "NVIS" approach. Since these radio waves are longer (literally 40–160M long), they reflect off the ionosphere better during certain hours. But let's back up for a minute.

Note: If you don't like technical details, you can skip this section and jump to the Pros, Cons, and Recommendations sections at the end of the chapter. However, if you're interested in learning more, keep reading!

Let's get the jargon out of the way. NVIS stands for "Near Vertical Incidence Skywave." It is also commonly described as "cloud-burning." Why? Every antenna transmits in a certain direction or multiple directions. In the case of an antenna positioned for NVIS, the radio waves

FIGURE 13-1: The Icom IC-7000. With the quarter as a reference, you can see that this HF/UHF/VHF radio comes in a relatively small package. It has plenty of power, numerous high-end features, and is used either as a mobile (vehicle) or desktop radio.

are directed straight upward. And when they hit the ionosphere (an atmospheric layer surrounding the Earth), they are reflected straight back down. These waves coming back down result in a signal covering a circle about 400–600 miles in diameter. This means you will be able to transmit on these frequencies and be heard by people up to a couple of hundred miles away (the radius). Of course, if they have a similar radio setup, you will be able to hear them too, when they reply. There you have it – regional communications!

As you can see in the following image (Figure 13-2), the radiation pattern of an NVIS antenna (the gray, shaded oval area) is essentially straight upward, and when the radio waves bounce straight back down, it creates a relatively small (for HF), regional coverage pattern.

For some types of HF antennas, instead of a lobe pointing at 90 degrees, as in Figure 13-2, you would see two lobes pointing in opposing directions, toward 150 and 30 degrees. Those patterns would result in

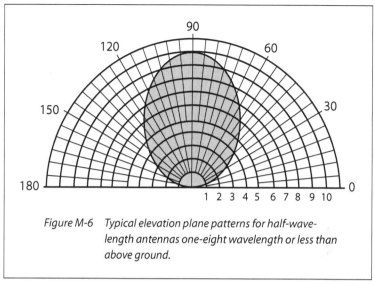

Figure M-6   Typical elevation plane patterns for half-wave-
length antennas one-eight wavelength or less than
above ground.

FIGURE 13-2: (U.S. Army field manual) NVIS radio wave pattern, aiming straight upward.

long-range signal transmission, because they bounce farther away at those steep angles.

For **local** communications, you can use UHF & VHF for up to several dozen miles with a repeater or a directional antenna with fancy names like "Yagi" or "log periodic." Directional antennas focus the radio waves in a specific direction and receive more effectively from the same direction.

For **regional** communications, you can HF and NVIS. For **long-range** communications, you can use HF with a "normal" antenna, e.g., a dipole, as long as it's placed high enough to direct those waves more horizontally than just vertically. There are many books on antennas that go into exhaustive detail, if you're interested in learning more.

Since most emergency scenarios do not realistically involve needing or getting quick assistance from hundreds or even thousands of miles away, we are not going to discuss long-range amateur communications in this book.

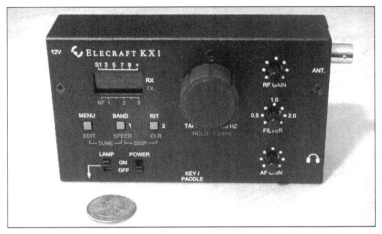

FIGURE 13-3: Sometimes touted as a "perfect survival radio," the Elecraft KX-1 can receive AM, FM, shortwave, and ham frequencies, and just about anything in between. It can transmit on multiple HF ham bands using Morse code. As you can see, it's very compact. And with a handful of AA batteries and a wire antenna strung up over a tree branch, you can transmit regionally or long-distance.

Why Get a General License?

The material you need to understand to get a General license is more complex and requires quite a bit more study, unless you are an engineer or already versed in electronics and RF theory. If you are not technical at heart, you may have to do some rote memorization. If you're like me, you may not even fully understand some of the answers you memorize. That's OK. Don't worry about it. While it's critical you understand some basics (e.g., don't let your antenna get close to a power line, don't transmit with 500 watts two feet away from your office desk, don't use your antenna as a lightning rod, etc.), you don't need to understand *everything* covered on the test. Of course, you will need to understand the FCC regulations and safety basics, but you will be able to learn much of the operating basics with the help of a local ham (or a mentor, often referred to as an "Elmer"), or by doing a little research.

Once you have your General license, get an HF desktop radio and set up a simple dipole antenna. If you position your antenna low enough, from five to ten feet off the ground (which is usually much easier to do than getting it up in the air), you will be able to do NVIS right off the bat.[21] Learn what works and what doesn't in your area at which times. Find a friend (or a whole batch of them) whom you can check in with regularly, and get familiar with the jargon, your radio, your antenna, your power supply, etc.

Please note: as I've stated earlier, HF NVIS antennas need to be low to the ground. UHF and VHF antennas will usually perform better high in the air. Don't mix them up.

The next level of licensing, "Amateur Extra," is far more difficult (it took me much longer), especially if you don't have an engineering background. It requires a lot more study of RF theory and other electrical engineering-related material, and many ham radio operators never bother.

To summarize, a Technician license is required, a General license is a great idea (and necessary for long-term, repeater-free, regional communications), and an Extra license is not necessary for most people.

Pros & Cons

Pros:

- With an HF radio, you can communicate across hundreds or thousands of miles, and in the emcomm context, HF with an NVIS antenna is a good solution for regional communications.
- HF can be operated at very low power, especially if you have a good antenna.
- Now you have a reason to learn Morse code! (Don't worry — it isn't required.) Morse code, also known by ham operators as "CW" from the term "Continuous Wave," is the first digital mode, and it doesn't even require a computer. Because of the tone used and its simple on/off nature, CW can be much easier to hear than a voice transmission, especially when a frequency band is experiencing atmospheric interference or other noise.

---

[21] 40M and 80M are easier bands to use. One reason is that their antennas are shorter (often 65 and 130 feet long, respectively) and therefore easier to set up. A common antenna length for the 160M band is about 250 feet, which is less convenient for many operators.

Cons:

- Almost all HF frequencies require the operator to get a General license from the FCC (one step up from Technician). Transmitting on any frequency that will also works with NVIS requires a General license.
- NVIS only works during certain times of day, depending on the band. This means that some frequencies will only be available during the daytime, but an entirely different set may be available during certain hours of the evening.
- Using HF effectively usually requires a better understanding of radio theory and antennas (hence the additional license requirements), as well as additional experimentation and practice with someone in your region.
- HF radios generally cost more than simple UHF/VHF radios, although you can find plenty of less expensive, used radios capable of meeting your needs. As with most hobbies, if you want to get hard-core, you can spend thousands on high-end radios, amplifiers, tuners, antenna rotators, and the list goes on. The good news: you can often find a decent, practical HF radio at a ham show or swap meet for $150-$300, and it will work just fine for emcomm purposes.

Recommendations

I recommend you get an FCC General license. It is not much more work than a Technician license and the additional information is worth learning. Find an inexpensive HF radio, set up an NVIS antenna and start talking with someone in your area.

This may mean you need to get help from someone in your local ham club or emcomm group. That's normal. Get out there and let the hands-on training begin.

CHAPTER 14:

THE SECRET TO GETTING AN FCC AMATEUR RADIO LICENSE IN RECORD TIME

Bob always wanted to be able to use cool radio gear, but was worried about getting an FCC license. A couple of his friends were into ham radio, but they had electrical engineering degrees, while Bob hadn't even gone to college. It seemed intimidating, and surely it had to be complicated and expensive, because everything involving the government was that way, right?

Then one of Bob's friends loaned him a book, "Personal Emergency Communications." Bob quickly read his way to the chapter on getting an amateur radio license. It actually sounded easy....

Bob decided to get it done. He followed the simple guidelines from the book, studied a bit, took a couple practice tests, and thought he was ready.

"I should have done this years ago!" Bob said to himself, as he filled out his application form. Then the test started.

After quickly answering thirty five questions (the answers were routine, after all his practicing) and double-checking his answers, Bob was done. But he wasn't the first. A teenage girl finished about a minute before him. "Kids do this, too?" he wondered.

Two weeks later, Bob proudly pinned his new license to his bulletin board. He picked up his new radio, already set to a specific VHF frequency, and started discussing his antenna setup with one of his buddies.

Getting Your Ham License Is Easier than You Think

If you want to be able to use the radio gear the amateur radio operators get to use, participate in a local emergency communications team, take your own emcomm planning and preparation to the next level, or do some of the other cool things hams get to do, you will need an FCC license.

What is the secret to getting your license? Is it understanding all of the amateur radio testing material put out by the FCC? Is it knowing every question and answer inside and out? Is it having a firm understanding of RF (Radio Frequency) theory? No, it is none of those things. Here is the secret:

Using the approach you like best, study the questions and the *correct* answers until more than 75 percent of them stick.

Understanding the why or the how for every question will not matter at test time if you don't provide the correct answer. Having a PhD in electrical engineering or a deep-seated desire to understand how radios work also isn't necessary. This test does not measure IQ or aptitude. It measures your ability to provide the correct answers. What matters is that you are able to produce enough right answers when you take the test.

Many hams (me included) learned most of what they know about radios and antennas by doing — setting up and using their gear, experimenting and working on projects with other hams *after* they got their licenses.

The test certainly isn't useless. One of the benefits is that all licensed hams should have a decent understanding of the FCC rules. This is better for everyone, because the ham radio airwaves are more civilized when we all follow them. It also does touch on many areas that do matter. Nonetheless, don't assume that studying for the test is a great way to learn all you need to know about amateur radio. Do you remember your driving test? It may have forced you to learn how to parallel park, but did it prepare you for the monster truck changing lanes on top of you, the crazed teenager passing you on the curved road at night, or your harrowing drive home after an unexpected snowstorm? Your driver's license was a ticket to the show, a requirement to participate, and never designed to bring you to the competence level that years of driving does.

Another way to view the test is as a gating mechanism. If you do not have a desire to use ham radios, you will probably not go to the trouble to take the test, regardless of its simplicity. The licensing process separates the genuinely interested wheat from the noise-making, rabble-rousing chaff, at least in most cases.

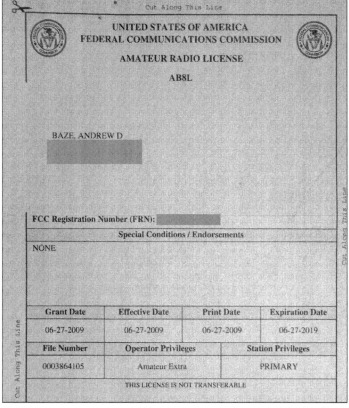

FIGURE 14-1: My most recent FCC license – an open door to worldwide, off-grid communications. You can easily get one of your own. Go do it!

There is one more thing you should know about getting an amateur radio license: there are no age restrictions. Go to any ham radio club meeting and it will be immediately evident that there are no upper age limits. But what about young people? Is there a minimum required age? Not that I'm aware of. I know there have been five-year-olds who have passed their test. That might give you some confidence. With a little study, it is obviously possible.

Here is the quick version of how to get your license:

1. Study all questions and answers ahead of time, using a book, flash-cards, CD or whatever else works well for you.
2. Take as many free, online practice exams as you want.
3. Take the real exam at your local ham club.
4. Wait a couple of weeks and get your license in the mail from the FCC.

"Oh no! It must be very difficult jumping through all the hoops to get such a license," you might say. Or you might wonder, "How could I possibly learn all of those technical concepts without an electrical engineering degree?" Or perhaps you're concerned about the astronomical licensing fees. Good news – none of this need concern you. Keep reading. Licensing is much easier than you think.

The test is easy to pass, and the license application is very simple. In addition, the people who administer the test will enthusiastically help you with any questions you have about the application. (Note: They will not help you with answers during the test. You're on your own there.)

When it comes to the test content, the concepts are not complicated. The electronics information you'll need to know is simple. For example, have you heard of the metric system? You'll need to know about that. Are you capable of remembering some simple FCC rules? For instance, "You have to say your call sign at least once every ten minutes when you are talking on the radio." Memorize some rules, some numbers, letters and other details that matter to the FCC, and you are just about ready.

I certainly don't mean to lead you astray. Of course, there are many more details you'll need to review, and unless you're a radio buff already, you will have to sit down and do at least a little studying and memorizing. But as I said earlier, it is not critical that you possess a deep understanding of every answer you provide. You will probably do most of the real learning when you are practicing using your radio, volunteering at a local event, talking with friends or members of your local emcomm group, etc.

There is one more very important fact you should know. *Every possible question and answer that might appear on the exam is already published and available for you to study.*

How many questions are on the test? 100? 200? Here is more good news. From the FCC website: "To pass the Technician Class examination, at least 26 questions from a 35 question written examination must be answered correctly." While they'll give you more than 35 questions and answers as part of the overall question pool from which to study, the test will only have 35.

I hear about people who don't pass the Technician exam. I know that in many cases the reason for this was:

1. Poor test-taking skills (e.g., not paying attention to which letter they're filling in on the answer sheet, even when they knew the right answer), or
2. Insufficient or inefficient studying.

I want to make this as easy for you as possible, to help you pass the test, get your license and open the door to all the options, power and fun ham radio can bring. Following are some easy-to-use tips. Read them and make your license exam far easier to pass.

Start by figuring out how you learn best. Since you already know that all of the possible questions and corresponding answers are available, you will need to decide the best way to review them so they'll stick in your head.

- Do you learn by **reading** and remembering? There are a few good books available that contain all of the questions and answers, as well as additional supporting info to give context to the answers. The additional info is especially useful if you learn better when you can understand the why of a new concept, or if you simply enjoy learning for learning's sake.

- Do you do better with **flashcards**? Those are also available.
- Do you learn most when listening? There are Q&A CDs you can listen to as you drive or relax at home.
- Maybe you are like me and you learn how to pass the test by **taking practice tests**. There are some great, free websites available for that. Alternatively, if you like, you can pay someone to access their online test-taking tools, which track your progress. (Links are shown later in this chapter and in the appendix.)

As I mentioned earlier, here is a tip you can use on **any** test that provides you with the questions and answers in advance: *Only read or highlight the correct answers in your study guide.* Cross out the wrong answers and only review the right answers. After you go through all of the questions, however many times you like, you will only have the correct answers imprinted on your brain, and when it comes time to take the test, you will only remember the correct answers!

Ham Radio License Exam Testing Resources

Here are some of the most common and useful books (most are reviewed on www.HamRadioBooks.com, if you would like more details):

- From the ARRL, with CD included: "**Ham Radio License Manual**." All the questions and answers are here, with some additional basic explanations — all you need to pass the test.
- From Gordon West, probably the biggest name in Amateur Radio license training: "**Technician Class 2010–2014.**" I like Gordon ("Gordo") West's books. He's old school, and takes a fun, interesting approach to learning about ham radio and getting the test out of the way. All the questions and answers are here also.
- Flashcards are available, if you learn well this way: "**Technician Class Flash Card Set**," from ARRL. I have not used these personally, but if you like flash cards, this set is probably your best bet.

- Here are some online testing resources. You can take the tests in advance as many times as you want for free! (There are other sites that charge for their services, but take a look at your free options first):
  - www.eham.net/exams/
  - hamexam.org/
  - www.hamtesting.com/index.php

Note: If you have only taken the time to read one or two chapters of a licensing book, take a free online test, and then find you only get 15 percent correct, don't get frustrated and give up! You just need to review the rest of the material in the license book. If you go through **all** the questions and answers before you take an online test, you will get much better results. This may sound like common sense, but I know folks who were so itchy to take the test that they failed to review all the questions and correct answers first, and then wondered why they didn't get all of the answers right. You know better.

What did I do? I started by reading through both of the Q&A books I mentioned previously. (Note: I often take an overkill approach. One book should be enough for most people.) Then, because I was curious about how ham radio worked, I read a book that explained additional details for beginners. Once I had a little foundation, I took several online practice tests until I could reliably score better than 80 percent. After that, I focused on areas where the answers made the least sense to me (even though I had crossed out the wrong answers in my book and was only focusing on the correct answers), practicing those a few more times, making sure I had the right answers fresh in my memory. For me, this plan made passing the test a piece of cake.

This was a more thorough approach than I'm recommending to you, because I mistakenly thought at the time that I would learn more about the most important aspects of using my radio by the extra studying. I found that in practice, on the job training was a far more practical use of my time when it came to understanding what mattered most.

This was my approach. Yours will probably be different. Find the study approach that works well for you, and when you're ready, take a couple of practice exams. Once you can reliably score at least 80 percent correct, you should be ready to take a real exam.

You can do it! Seriously, how often does a test already have all of the answers available to you in advance? Get this non-obstacle out of the way and add one of the best tools available to your emcomm toolbox!

Where to Take a Ham Radio License Exam

Finding a place to take the exam should not be difficult. When I looked, I had a hard time finding a place, but I was on my own with nobody available to help. The following resources will make it much easier for you.

- The ARRL: http://www.arrl.org/find-an-amateur-radio-license-exam-session. By the way, after you get your license, I recommend you join the ARRL. Go to www.arrl.org and find the "join" link. Get your subscription to **QST Magazine**, full of great articles on amateur radio, delivered to your mailbox every month. The ARRL helps keep amateur radio frequencies available to us all, and the magazine is a great way to learn more about the hobby, whether you are a beginner or expert.
- www.hamdepot.com
- www.dxzone.com/catalog/Ham_Radio/Clubs/

That's it. You can review all the answers at your leisure, and find a club nearby that will let you take the test and help you (if you need it) with the simple FCC form.

Ham Radio FCC License Fees

You're not done yet. I forgot to tell you about the very expensive license fees…

An FCC amateur radio license is **FREE**. You may have to pay $10 or $15 to take the test, which helps the club running the test pay for their supplies, room rental, and postage to ship off your answer sheets to the FCC, etc., but the cost of the license itself is indeed free.

But wait, what about when you renew, ten years later? (Yes, every ten years. Pretty convenient, right?) Is that is when you have to pay? No, it's still free. However, there is a special circumstance when you *could* pay, if you really wanted to. If you want a **vanity call sign** (where you pick the letters and/or numbers you want, if they are not already in use), you will have to pay. Some hams get a call sign that includes their initials, like me – my call sign is **AB8L**. Is that expensive? I'm sorry to disappoint you again. At the time of this writing, it will cost you a whopping $14.25 for a vanity call sign. Renewal ten years later is another $14.25. Even the least indulgent among us is not likely to balk at the "vain" expense of $1.43 per year.

As I touched on previously, there are currently three license types in amateur radio:

1. Technician – this allows you access to a specific set of UHF, VHF, and some HF frequencies, which will give you the ability to do short- and some medium-range communicating.

2. General – this allows more access to other frequencies, including much but not all of the HF amateur radio spectrum, which is important for medium- and long-range communications (including NVIS communications, important for regional emcomm as explained earlier).

   Tip: if you are studying to take the Technician exam, you may want to take the General exam at the same time. You can take more than one test in the same testing session for no additional charge!

3. Amateur Extra – this allows access to all amateur radio frequencies. You will most likely need to study much harder for the Extra exam, because it is far more in-depth and requires much more knowledge of electronics, radio wave theory, etc. But Technician and General are probably all you will ever need.

Now go out and get a license, at least Technician level, and I hope to hear you on the air soon.

MORE IMPORTANT TOOLS

By now, you have answers to key emergency planning and technology questions, your own personal plan, a calling clock, and you may even have some cool, new gear in your hands. But don't stop reading yet. There are other important considerations to keep in mind, which are covered in the next chapters.

By getting this far, you are able to be more prepared for emergency communications than 98 percent of the population. Now let's tie up some loose ends and bump that number up to 99!

CHAPTER 15:
POWER – THE KEY!

You may have removed your dependency on the Internet and phones, but what about the AC outlet in your wall? Don't forget that you'll need a power supply for your communications devices, with the exception of a hand-crank radio.

The two most common power supplies are:

1. 120-volt (120V) AC (Alternating Current) power, which is what comes from the outlet on your wall (your "grid power" supply), a generator or an inverter.
2. 12-volt (12V) DC (Direct Current), from an AC transformer, some generators, a battery or a battery pack.

For radio communications purposes, AC power is usually converted to DC, which the radio consumes, or uses to recharge the radio's battery pack. In an emergency, you should assume that you will not have 120V AC grid power, although you may have a generator available to provide temporary power.

FIGURE 15-1: 12-volt power plug. Does your radio have one? With no grid power available, 12V batteries are a superior alternate power solution.

DC power is usually supplied via batteries. The alternative is a transformer, which transforms AC to DC, e.g., the "wall-wart" you plug into a wall to power certain electronics. Other DC power sources are the 12V battery in your vehicle, a deep-cycle 12V battery in your boat, a 6V golf cart battery, the 9V battery in your smoke detector or a common 1.5V AA battery. These all provide DC power.

For simplicity's sake, let's assume you don't have a generator or any other AC power source available after a serious emergency. Even if you do, at some point you'll need batteries.

One very useful option is to keep a 12V deep-cycle battery charged and ready. Why? When properly configured (with a cable and a 12V outlet), you can plug in any device that uses a car charger.

FIGURE 15-2: This is a 12V outlet, with an Anderson Powerpole plug on the opposite end of the cord. The Powerpole plug can connect this outlet to a battery, solar panel, or any other 12V source.

This book doesn't cover setting up a complete backup power system. For now, you'll need to do your own research, but keeping a 12V battery ready to use is a great idea.

A simpler solution that you should definitely consider is to have a large package of AA batteries handy. (Of course, if you have gear that uses AAA batteries, you'll need to get packages of those too.)

If you have a "go bag," "bug-out bag," or "get home bag," you should probably invest in some lithium AA batteries. Why?

FIGURE 15-3: 32 amp hour AGM battery made by Deka. This battery is capable of recharging many portable devices many times over.

Pros & Cons

Pros:

- For high-drain electronics (which your radio probably qualifies as), lithium batteries will last longer. For example, the packages advertise "8x as long" with digital cameras. They may not last quite that long for you, but they should last noticeably longer than standard alkaline batteries in a radio.
  - Note: The lithium disposable batteries I'm talking about are *not* the same as lithium-ion rechargeable battery technology (which many rechargeable batteries use).
- Lithium disposables are much lighter than alkaline batteries, which is important if you need to carry them around.
- They store much longer than alkaline batteries. Energizer® advertises a lithium battery shelf life of up to fifteen years. Check the expiration dates on battery packages and you'll see for yourself. I have not personally tested any 15-year-old lithium AA batteries (and I'm not sure the product has been available for more than a few years), but I trust that they last much longer than my alkaline batteries. I have encountered unused alkaline batteries that were several years old and didn't work.
- If you need to expose your batteries or electronics to temperature extremes, lithium batteries are said to function in temperatures from -40°F to 140°F. Alkaline batteries will not work well in such extremes. (I have personally experienced significant decrease in alkaline AA performance in temperatures below 32°F. Lithium batteries should work in temperatures far colder.)

Cons:

- They are more expensive than alkaline batteries. Amazon.com prices are decent, and probably about the same as you would pay for regular alkaline batteries at your local Quick-e-Mart.

Recommendations

Buy at least one or two packages of lithium A or AAA batteries, as needed for long-term storage, especially if you need to carry any spares with you.

Ideally, buy at least enough to run all of your critical portable electronic equipment for three to seven days. This should not add up to a large expense or storage space. You will have to spend a little bit, but this is a long-term storage resource, and you will use them eventually.

FIGURE 15-4: Energizer Lithium AA batteries. They are lighter than alkaline batteries, store more power, and will stay fresh for up to 15 years.

If you're on a tight budget and know you'll be able to cycle through your batteries without wasting them by letting them expire, go ahead and stockpile some alkaline ones.

Another good idea mentioned briefly earlier, though a little more complicated to set up and maintain, is to get a deep-cycle 12V battery with a trickle charger/battery maintainer and the necessary adapter(s) to allow you to charge your important electronics. In addition, you should get an inverter, which will convert 12V DC power to 120V AC power. This will allow you to run AC-powered devices directly from your battery.

For more information on backup power solutions, I strongly suggest you read the ARRL book, *"Emergency Power for Radio Communications."* It is a fantastic book for beginners and experienced radio operators alike. It provides a wealth of information on backup power solutions for radio, lighting, running your refrigerator or furnace, or whatever else you might need to power in an emergency. You can also find a detailed review of this book at www.HamRadioBooks.com.

CHAPTER 16:

BECOMING MORE SKILLED WITH YOUR GEAR

You now have a plan. You've made the effort to evaluate your situation and your needs. You've decided what you need and then acquired at least the minimum necessary set of gear. But learning more probably caused you to ask many more questions. What's the best way to find the answers to those questions?

Internet forums can be a great way to learn more. But beware. Not all information sources are equal. You will get much more value from the people with firsthand experience than from those who can only speculate or regurgitate whatever they've heard.

As with any pastime, you can find a subgroup of people with one behavioral pattern in common: they *talk*, but they don't *do*. Not all people you come across will be that way, but it's a good idea to pay attention to your information sources.

If you visit nearly any hobby- or activity-focused Internet forum, you can usually find two general categories of people (in addition to the "lurkers" who read but don't actively participate):

1. Those I just mentioned, who talk about what's "clearly best" to do or have, regardless of whether they have ever done the activity or used the equipment themselves. They heard something from someone at some point, or read a comment in one of their favorite blogs, and are repeating it as "fact." They are often noisy and confident in their opinions.

2. Those who really have "been there and done that." Their comments are based on personal experiences. *These comments are the ones that matter.*

How do you determine which comments are worth paying attention to? Ask a simple question, whether in person or online, like this: "How do you know?" Of course, you'll need to ask it appropriately. For example, if someone says, "This radio has the best range, thirty-two miles" (a loaded statement in the first place, because it contains the word "best"), you can reply with "Wow, that's great. Have you used it at that distance, or do you know someone who has? I would love to hear about how the sound quality was at that range."

Ideally, you are not directly challenging the person who made the comment, because many people will get defensive, especially those without any hands-on experience on the topic they're discussing. Even if the person has no direct experience, he or she may be able to refer you to someone who does, so you don't want to burn that bridge outright. Your goal in this case is not to alienate a potential resource, but to obtain useful, realistic information.

Let's take it one step further, since we're talking about opinion and experience. Now that you've gone to the trouble to make an emergency communication plan, are you ready to put your safety, or the safety of your family and friends, in *my* hands?

You might want to read that last question again and give it some thought.

Until you use your equipment and try everything yourself, and even though you value my opinion enough to read this book, I don't want you to rely solely on my opinion and experience. My recommendations may not be exactly what you need. You need the confidence that comes from firsthand experience. You need to know personally that your gear works, that the frequencies you have selected will not be blocked by interference, and that your rechargeable and disposable batteries will last the expected amount of time. This means you need to get out and *use your equipment.*

Get Out!

Many people have a hard time practicing with their gear for practice's sake. I'm this way myself. I'm not motivated to practice with radio gear unless I have some other reason to use it.

My approach to forcing myself to use my gear, as well as meeting my other personal goals, is to participate in groups that use radio gear as part of their regular operation.

For example, I am part of my local ARES group and other local emergency communications teams. As part of my active participation, I check in each week on their radio net.[22] It only takes a couple of minutes, but because I do, I now know some important things:

1. How to change frequencies correctly, especially after I accidentally fiddle with a dial and mess up the frequency I was just on, right before I am supposed to be talking or even running the net. (Oops!)
2. How long my batteries last.
3. The general range I can expect on different bands in different terrain, and which locations will allow me to reach various local repeaters.
4. What my signal will sound like (very poor to very strong) in various locations, using different antennas.
5. Many other minor yet important details.

---

[22] A radio net is a common way to facilitate group radio communication. A "net control operator" usually runs the net. A net script (usually read verbatim to maintain consistency as well as to help calm public speaking nerves) commonly follows a format like this one:
  1. Brief explanation of the net's purpose and format
  2. Calls for participants by name/call sign, and whether they have any announcements (aka "traffic")
  3. Detailed announcements
  4. Another call for late check-ins or other news
  5. Close
Without this organized approach, it would be impossible for a group of people to communicate effectively.

I probably would not know the vast majority of this vital information if I hadn't participated actively in these groups. And at the same time, I got to learn a lot about emergency communications processes, which certainly didn't hurt.

Please note that in the last few paragraphs I mentioned "active participation." This is an important distinction to make. "Inactive participation" usually means simply listening to a weekly radio net and maybe coming to a meeting only when it is absolutely required. That's still possible in many cases, since volunteer emcomm organizations are usually lightly staffed and looking for anyone they can get to help; however, you will get vastly more learning, experience, and other value out of active participation.

Do you think you could read a book about horseshoeing (even an exhaustive one with lots of pictures) and be able to step up and competently shoe a horse? I don't think I could. Therefore, don't expect that several hours of book learning (or simply passing a licensing exam) will prepare you to use your radio in everyday life, much less in the stressful conditions of an emergency. You will get far more out of doing than observing.

If public emergency communications aren't interesting to you, fear not. After all, this book is about putting together a *personal* plan. There are myriad other options available for getting the basic experience and confidence necessary to run your equipment in a pinch. Do you like parades? I'm not joking. Many parade organizers tap amateur radio volunteers, placing them at various locations as information gatherers or as a way to relay information. Other events, such as marathons, bicycle races, or other large public gatherings, regularly use licensed, ham radio-toting volunteers as a communication backbone.

How can you find out what's going on in your area? Your local ham club will have information on what's planned. I wholeheartedly encourage you to identify something you find particularly interesting and then find a way to volunteer your unique skills in a way that also lets you exercise parts of your emergency communication plan, as well as become much more familiar with your equipment in ways you may not otherwise do.

Remember, simply reading the manual (if you can make it through one of those dry tomes) or someone's opinions or experiences on an

Internet forum is no match for your own personal experience. Go get some experience!

In Washington State, here is list of just some of the events currently using ham radio volunteers:

- March of Dimes March for Babies, Tacoma
- Capitol City Marathon, Olympia
- Walk MS: Seattle, Husky Stadium
- WA Special Olympics, Joint Base Lewis McChord
- RAMROD Bike Ride, Enumclaw
- Seafair Torchlight Parade, Seattle
- Danskin Triathlon, Seattle
- Veterans Day Parade, Auburn
- Seattle Marathon, Seattle

If your closest town or city has any organized bike races, foot races, parades, walks for charity or other large events run by volunteers, the odds are good that you will have an opportunity to help and practice your radio skills at the same time.

An Important Note – How to Talk on the Radio

If you are new to using a handheld radio, you may benefit from a couple of tips on how to speak effectively on the radio. It is not a telephone and requires a slightly different approach to communicating clearly, interrupting appropriately, coordinating communications with large groups, dealing with emergencies, etc. Keep these tips in mind:

- If you need to interrupt, as soon as there is a pause in an ongoing conversation (and if they are not dealing with their own, equally important emergency), interject with, "Break, break." Good radio operators will pause to say, "Go ahead."
- Be succinct in your question or request. Do not suck up precious airtime, especially in an emergency. Get right to the point and describe exactly what you need.

- Speak clearly and in a normal tone of voice. Raising your voice will not make your signal any more powerful. In fact, in some cases it will decrease your readability. (By readability, I mean "listen-ability," how well others can hear you.)

- Speak across your microphone, or at a slight angle, versus directly at it. It may not be intuitive, but this approach will allow the microphone to more effectively pick up your voice, without the minor distortions caused by the puffs of air (like when you pronounce a "p" or "t"), to which microphones are often sensitive.

- Do you plan to be in a noisy environment with a need to speak over competing sounds? Or will you be in an area where you don't want to disturb others? In both cases, using a headset with your radio can reduce the noise heard by others and increase your clarity.

CHAPTER 17:
YOUR "GO KIT"

If you don't have a three-day emergency kit, you'll get even more value from this chapter. While the focus here is on incorporating communications into your emergency kit, we will briefly cover most of the contents you will need.

There are many types of bags and kits out there, from the "Get Out of Dodge" bag (GOOD), the "Get-Home Bag" (GHB), the "Bug-Out Bag" (BOB) and several other creative variations. Gear-heads love discussing these variations and their personal setups, as well as the scenarios in which they could be used.

This is our general scenario: The power is out and you have nothing but this bag. You want to execute your emergency communication plan and contact the people on your list. At the same time, you want to be safe and relatively comfortable.

Nowadays, the three-day emergency kit is much more in vogue, and available at Costco, Wal-Mart, Amazon.com, Fred Meyer, and many other popular sources. It may be easier to add certain communication and power equipment to an existing bag, versus creating a completely new bag from scratch. But be careful. Most contain some or are even made up entirely of low quality gear. Make sure you know what is inside before you spend your hard-earned money on a commercial bag.

Even with high quality commercial bags, you must look over all of the contents and be prepared to replace some of them. Make sure the bag fits your unique needs, because "one size fits all" usually will not fit you. If you want the kit that's best for you or your family, you can save money by putting your own together. It is not complicated.

The following lists are to get your creative juices flowing, and are not meant to be an exhaustive solution for your situation. I have no idea what you'll need, although many situations will require some or much of the gear listed. You still need to invest the thought and effort to equip yourself appropriately.

Let's quickly review how people usually carry their loot.

1. Everyday carry (EDC): this is the gear people regularly carry with them, whether on a belt, in pockets, or in a purse or "murse," (aka "man-purse") which they always have with them.

2. Other bag(s): Carry-all, tote, backpack, range bag, get-home bag, briefcase, computer bag, bug-out-bag, survival kit bag, or any variation thereof.

Let's get the EDC basics out of the way, and then we'll dive into what goes into your radio kit and any other emergency bag you will have available.

Everyday Carry

While the EDC topic is fascinating to some, we will only review it briefly. If you want to dive headfirst into the world of unique multi-tools, titanium toothpicks, cutting-edge flashlights, other cutting edges and gadgets, visit www.edcforums.com. Usually, anything that might be conceivably carried every day has been discussed at great length on this forum.

Here are my very brief and most basic recommendations. Aside from the normal junk you need to manage your life, (e.g., wallet, keys, lip balm), you should also think about carrying these:

1. **Pocketknife**. A knife is an incredibly useful tool, and I have carried one for most of my adult life. Once you carry one, you will realize how many uses it has. If you don't, I suggest you start.

2. **Multi-tool**. This may seem extreme to many, but today's multi-tools come in all sizes and contain any manner of handy tools, from the common screwdriver to bottle opener, saw, file, pliers, scissors, ruler, wire cutters, even USB-compatible thumb drive storage. They can fit on a belt pouch, in a purse, or on a keychain.

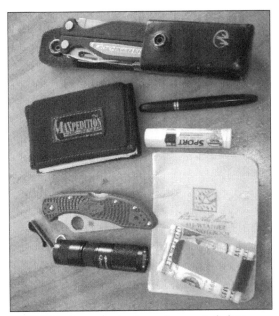

FIGURE 17-1: One example of everyday carry includes a Leatherman Charge multi-tool (top), wallet, pen, lip balm, Spyderco Delica pocketknife (middle left), 4Sevens flashlight (bottom left), cash, and small all-weather notebook.

3. **Flashlight.** With today's battery and LED technology, there is no good reason to not carry at least a small LED flashlight. Some are no bigger in diameter than a quarter and use lithium batteries, which will often last for years with light use (no pun intended). If you ever lose power or need to go where there is no artificial lighting, the odds are 50-50 as to whether you'll be able to see. Walk around for a minute or two with your eyes closed and consider those odds without the distraction of your primary sense of sight.[23] Then go get a good light and make those odds 100-0.

---

[23] Are you legally blind, listening to this book on your e-reader? Or are you helping plan for someone blind? My recommendation to you is the same. Carry a small flashlight anyway. If you have an emergency at night, don't have natural lighting available, and can't use your other senses to solve the problem on your own, that light can certainly help others to help you.

I could write an entire chapter (and probably an entire book) on this topic, but since we're focused on communications in this case, I will leave you with that guidance.

Now let's take a look at what should go in a communications-specific kit.

Radio Kit

Note: this list is intended to be for everyday people as part of a personal emergency communication kit, not for an ARES/RACES or other Emcomm team member to support his/her team or local government. That person's go-bag should include many additional items (e.g., binder containing key documentation and forms). In addition, such a team should have detailed go-bag guidelines or requirements available.

Here is a basic list of what you will need in the communications section of your go-bag:

- A copy of your Personal Emergency Communications Plan
- Depending on what's in your plan, any combination of the items listed below:
  - FRS/GMRS radio
  - AM/FM/shortwave/weather radio
  - Scanner
  - Handheld amateur radio (HT)
  - Other device (although the above are most common)
- AC and DC (car) radio power supply/charger(s)
- Backup AA batteries (see Chapter 15 – Power), one or two sets of batteries for every device
- Radio manual or "Nifty!" guide,[24] as needed

---

[24] Most amateur radio manuals are long and somewhat difficult to easily use as a reference. One man created a business out of selling rewritten, summarized (in some cases), laminated, spiral-bound manuals that humans can read and easily use. Bernie Lafreniere (call sign N6FN) has painstakingly compiled readable manuals for most popular ham radio models, and calls them "Nifty!" guides. He also put together small, folding, laminated, quick-reference cards, which are very handy to have in your radio's case or even attached to the back of the radio with a rubber band. You can find more information at www.NiftyAccessories.com.

- Small LED flashlight
- For amateur radios, a few additional accessories should be considered, which will add important flexibility or range:
  - Aftermarket antenna (not the inefficient "rubber duck")
  - A mag-mount antenna to allow the handheld radio to be used from within a vehicle
  - Speaker microphone
  - Headset
  - Adapters and jumper cables (usually used instead of a simple adapter, to reduce stress on antenna or coax connections)

Do you wonder what you'll see in the go-bag of an emcomm volunteer? They will take some of the areas we covered a step further and add items such as these:

- Larger 12V battery if space/weight capacity is available, with cable and 12V receptacle or Anderson Powerpole[25] plug
- Inverter (converts battery power to AC power)
- 12V cable with clips to attach to 12V car battery
- Powerpole adapters
- Rollup J-Pole antenna and coaxial cable with adapter that fits an HT
- Additional coaxial cable and power connector adapters
- Frequency and repeater directory
- Amateur radio license
- Emcomm ID cards
- Group roster, process documents, key forms
- Clipboard with plastic case (functions like a portable desk, extra pens, paperclips, etc.)

You will find other indispensable options. Include them as you see fit and feel free to send me an email about anything that appears to be missing from my list of basics.

---

[25] The Anderson Powerpole is a reliable, easy to install and use DC connector. It's a standard for many emergency communications groups.

The Bag(s)

Just to be thorough, let's briefly review the common items that you should consider for the rest of your go-bag.

- Food, snack bars, MRE (meal, ready to eat), snacks
- Water, a water filter or purifier and container
- Lighting – two LED flashlights, one of which can clip somewhere to provide task lighting
- Paper printouts of key contact information, addresses, phone numbers, etc.
- Local map
- Digital data – thumb drive, encrypted as needed
  - Personal documents
  - Radio manuals
  - Local maps
- Multi-tool and pocket knife
- Pepper spray, if needed
- Cell phone backup charger
- Mini survival kit
- 550 cord/rope
- Garbage bag and Ziploc® bags
- Lighter and a backup method to make fire
- Paper and pencil
- Map(s)
- Money (coins and paper)
- Alcohol-based hand sanitizer
- Toilet paper
- Extra clothing (depending on climate and season)
- Spare eyeglasses
- First aid kit
- Prescription medication
- Calling card
- Quarters (to use in pay phone or vending machine, if they work)

None of these lists are meant to be definitive. Again, think about your plan, your climate, your personal needs, the needs of others in your plan, the time of year (you may want to swap out components twice a year as seasons change), etc. Every person and preparation scenario is different.

CHAPTER 18:

CLOSE THE LOOP - EXERCISE YOUR PLAN

Let us assume that you're familiar with your gear and have fleshed out your plan with realistic detail. What remains?

You're not done yet! You must test your plan. If you have a plan you have never used or tested, it will have holes, possibly huge ones, and worst case, dangerous ones.

You need to walk through every step of your plan in two different and important ways:

1. A mock (also known as "table-top") exercise, in which you step through each line of your paper or digital document with an additional participant or observer.
2. An active test, using your documentation and equipment each step of the way, with at least one other person.

With the mock exercise, simply walking through your plan with the people who are part of it will usually identify critical gaps. An added bonus is that you'll get another point of view. Any other person's perspective is invaluable. A different set of eyes will see things your eyes may have missed. You'll quickly learn what your incorrect assumptions were, or what you simply forgot to document.

During this process, do not take any criticism or suggestions for improvement personally. Instead, you should view all comments as opportunities to learn and improve your plan. Take notes, because the next exercise will be more realistic and less tolerant of deficiencies or gaps!

Now that you've walked through your plan, you're off to a good start. But you're not quite done. A real-time test is next. Find someone else interested in testing your plan with you (ideally someone already called out in your plan), and give him or her a copy of the plan. The test begins with you saying, "We've just had a disaster, and the plan is now in effect. Go!"

Make sure you and your partner have pen and paper ready to take any notes. You'll need them.

Then you should separate by an appropriate distance and start using your document to communicate. You and your partner should both be able to read the document and figure out what to do and when to do it. See what happens. It's OK to use your cell phone as a backup tool, so that if part of your plan is missing a key component, you can move on and test the other parts.

Remember, if something in your test doesn't work as expected, this means your test was successful. A good test will help you find out what needs improvement. It's not often that a test shows everything to be perfect. If something doesn't work as you expect, write it down and add it to your "successful test" list.

How do you move from one section of the plan to another? Here is an example. If your cell phones work and you make contact (which is Plan A for many people), then at the end of your first contact, you can say, "OK, assume voice communications don't work anymore. Now what?" Whatever you have designated as Plan B should be the obvious next step. Keep introducing variations on the scenario as you work your way through the plan, making sure you cover each section of the plan. Part of your job, in addition to exercising the plan, will be to inject potential problems, forcing you to move from Plan A to Plan B to Plan C, etc.

Don't forget this part of the test: at some point in the plan, let your partner know, "The power is now out. Cell towers and radio repeaters will only work for another twenty minutes." If you have a radio repeater in your plan, I suggest you add, "Repeater X has been taken over for emergency services use only. We can't use it anymore." See what happens.

You get the idea. Keep throwing contingencies into the mix and see how your document addresses them. If you need to solve problems on the fly, that's fine, but don't forget to document those problems and the new solutions you've identified. If it's not clearly specified in your plan, write it down and add it to the plan later. The goal, however, is not to create fantastic, unrealistic scenarios (or to annoy your partner), so don't overdo it. The goal is to test the document as thoroughly and realistically as possible, and clearly determine whether your key scenarios are covered.

When you get together at the end of the exercise, you should walk through what happened, step by step, and note any additional changes you need to consider. This is important – don't skip this step, referred to as a "post-mortem" or "after-action review." Make sure you ask these key questions:

1. What went well?
2. What went poorly?
3. What needs to change?

Don't put this step off, because you'll probably forget some important details if you wait.

Pat Yourself on the Back!

Now look back at what you had when you started reading this book (especially if you had no plan) and compare it to what you have now.

Do you remember the original questions?

- How can you contact anyone if your landline phone, cell phone and Internet connection do not work?
- Will you be able to talk with family and friends after a serious emergency or disaster?
- Do you have a communications section in your emergency plan?
- Have you tested it?

If you followed the instructions I have laid out, you should have good answers for every question above. Pat yourself on the back – you deserve it! Then loan this book or buy another copy for anyone you know who could benefit from the information.

Thanks for reading and stay safe!

RECOMMENDATIONS ROUNDUP AND APPENDICES

In this section, I've listed the recommendations I covered in Chapters 1 through 18. In addition, I thought it might be helpful to add more resources, such as a list of the links I used in the book, more information on getting formal emcomm training, and additional planning templates for you to review.

CHAPTER 19:

RECOMMENDATIONS SUMMARY

To summarize, here are my recommendations for components of an em-comm plan that will accommodate most people. Must-have components are shown in **bold text**.

- ✓ Make a **written emcomm plan** (Chapters 1–3).
- ✓ Get an **AM/FM/SW radio with** NOAA alerting and multiple power sources (Chapter 5).
- ✓ Purchase a scanner and get familiar with local radio traffic (Chapter 6).
- ✓ Buy and practice with a **set of FRS/GMRS radios** (Chapter 7).
- ✓ If CB is prevalent in your area or otherwise a great fit for your plan, set up base station and/or mobile CB radios (Chapter 8).
- ✓ If you need secure, short-range communications, get a set of eXRS radios (Chapter 9).
- ✓ If you're out trekking on your own and want a simple solution to send a one-way message home, get a SPOT device (Chapter 10).
- ✓ If you can afford a satellite phone, find an appropriate minutes package and test it (Chapter 11).
- ✓ **Get an FCC amateur radio Technician license** (Chapter 14).
- ✓ Get a **UHF/VHF handheld amateur radio**, set it up with local repeater frequencies, and start using it (Chapter 12).
- ✓ If you need regional communications, get your FCC General license, an HF radio, and set up an NVIS antenna (Chapters 13 and 14).
- ✓ Set aside **batteries for emergency power** when grid power isn't available, even if you have a generator (Chapter 15).

✓ Get involved with a local ham club or emcomm group and **practice using your radio equipment** (Chapter 16).

✓ Set up your **emcomm "Go-kit,"** in addition to any emergency kit you may already have (Chapter 17).

Integrating the above **bolded** critical components into your emcomm plan will put you light years ahead of most of the population. Welcome to realistic emergency communications preparedness!

APPENDIX A:
ADDITIONAL RESOURCES

There are hundreds of great Internet web sites, blogs, and even books, chock full of useful information. In my experience, however, most of my best information has come from other hams, even information on non-ham-radio topics, such as FRS/GMRS and CB. Two-way voice communication isn't a hot topic unless it involves the latest cell phone technology. (E.g., do you have 4G or 5G network speed?) You will not see the latest FRS or amateur radio handheld discussed on popular sites like Engadget or CNET. And you may see ham radio mentioned once in a blue moon, usually in a feel-good news piece after some kind of emergency or disaster, when hams helped relief efforts.

Though I've seen some very interesting amateur radio-focused blogs and websites, many of them appear to use outdated design and aren't very popular in general. But there are a few exceptions:

- www.ARRL.org – the granddaddy organization for amateur radio (although it still gives critical attention to engaging youth!)
- www.Eham.net
- www.CQ-Amateur-Radio.com
- www.HamRadio.com – Ham Radio Outlet provides excellent customer service and competitive prices for amateur radio equipment and other radio gear, books and more.

I would be remiss if I didn't mention another site I run, dedicated to reviewing ham radio books:

- www.HamRadioBooks.com

For quick reference, here are other links I've mentioned in this book:

- www.EmergencyCommunicationsBlog.com
- www.SWLing.com
- www.weather.gov/nwr
- bizarrelabs.com/foxhole.htm
- www.satcure-focus.com/hobby/page6.htm
- www.crystalradio.net
- www.HomePatrol.com
- www.TriSquare.us
- www.eham.net/exams
- hamexam.org
- www.hamtesting.com
- www.arrl.org/finding-an-exam-session
- www.hamdepot.com.
- www.dxzone.com/catalog/Ham_Radio/Clubs
- www.ARRL.org/membership
- www.globalstar.com
- www.dxzone.com/catalog/Ham_Radio/ARES_RACES
- www.edcforums.com

If you only do one thing to connect to other ham radio enthusiasts, join the ARRL. The Amateur Radio Relay League is run by avid supporters of amateur radio. They educate young and old, lobby the government to preserve amateur radio frequencies, and much more. Membership is inexpensive and comes with a subscription to *QST* magazine, a great resource for learning more about all aspects of ham radio. Learn more and sign up today! As mentioned earlier, their address is www.ARRL.org/membership.

I sincerely hope you found this book to be helpful and informative. If you have any remaining questions, comments you would like to share, recommendations or great emcomm planning or gear examples, please let me know by posting on www.EmergencyCommunicationsBlog.com (or by sending me an email through the site).

APPENDIX B:

GETTING HARD CORE – EMCOMM TRAINING

You may find emergency communications very interesting, versus just another component of your personal emergency preparedness plan. If you feel drawn to helping your community or local government prepare for loss of traditional communications, you have options.

Earlier I referenced volunteering as part of your local ARES/RACES (or the closest approximation) team. These groups are often tied to a local government in some way, either through fire or police departments. In some more urban areas, the local government has a dedicated "emergency management" office that handles coordinating communications and government response. If you call up city hall and ask who manages emergency communications in a disaster, or who coordinates the city's emergency response plan, you should be off to a good start. Who knows? If your area doesn't have anything like this, you could set up the first ARES group and help your whole region be better prepared.

You may wonder what the difference is between an emergency communications plan for a person/family versus a local government or agency. When more people are involved, more detailed planning and standardized training are required.

Here is one kind of training. As part of my ARES team, a couple years ago I took the ARRL's "Amateur Radio Emergency Communications Course – Level 1." It was informative, and, although the curriculum is currently being revised, the general concepts will endure.

The curriculum contains a lot of what you would probably expect, and covers these topics and more:

- Emergency communications systems
- The Incident Command System (ICS) method of organization
- Communications skills
- How to participate in and run emergency nets (i.e., coordinating radio discussions in an emergency)
- Message handling (how to systematically document, relay and track important messages)
- How to effectively deploy (what should be in your emcomm go-bag, etc.)
- Equipment choices
- Personal safety, survival, and health considerations
- What to expect in a large-scale disaster

As you can see, going through such a course (two days long in my case) will give you a more firm grounding in realistic emergency communications practice, and the opportunity to use these skills as part of an emcomm group.

Aside from calling your local city hall, as I mentioned earlier, you may also find more information on existing ARES/RACES organizations in your area at this site: www.dxzone.com/catalog/Ham_Radio/ARES_RACES/.

APPENDIX C:

ADDITIONAL PERSONAL COMMUNICATION PLAN EXAMPLES

I demonstrated one detailed emergency communication plan example in Chapter 3. Your situation will likely be different and I'll walk you through a few more examples that may suit you better. In any case, you can go online and download an empty template or one of these examples to modify.

In the following example, we have two single friends who want to stay in touch. They only have one out-of-state contact. This plan is not very robust, but it is far better than nothing.

Two things to consider with this lightweight plan:

1. They have an out-of-area contact. This is a good thing and should be part of every plan.
2. Even though there are no amateur radios as part of this plan, they do have a section on how they will use their FRS/GMRS radios. Again, I strongly recommend everyone get a set of these to augment the grid-dependent cell phone or landline systems.

Plan A	Method	Cell phone
	Who	Mary and Susan call each other
	When	As soon as possible, after emergency happens
	Phone number	Mary's cell: 426-555-1212, Susan's cell: 426-555-2121

Plan A (cont.)	Frequency	NA
	Notes	If no answer, leave a voicemail, followed by text message. If no answer in two hours hour, proceed to Plan B. If phone is below 50% charge, shut it off and turn it back on for fifteen minutes, five minutes before the hour until ten minutes after the hour, every hour. If no signal is available after the first five minutes, turn the phone off and try again the next hour.
Plan B	Method	Cell phone and landline phone
	Who	Attempt to reach Ron (Susan's father)
	When	Every hour, on the hour, one call
	Where	Eastern Time – Maine
	Phone number	Ron Parker cell: 426-555-1212, landline: 207-555-2121
	Frequency	NA
	Notes	If answered, leave a message with details, including any alternate contact information for yourself. If no answer, leave a voicemail, followed by text message (to cell phone). Go to Plan C regardless.
Plan C	Method	FRS/GMRS radio
	Who	Mary and Susan
	When	Every hour for fifteen minutes, starting on the hour
	Where	Same time zone
	Phone number	NA

Plan C (cont.)	Frequency	Channel 16, privacy code 14 (the default setting for both radios)
	Notes	Fifteen minutes every hour, calling every 2–3 minutes. if sufficient backup battery power is available, other information may be learned by setting the radio to "scan" during the rest of the hour, and listening to nearby conversations.] If radio(s) are less than 50% charged or batteries are scarce, call only on even hours.
Other details	NOAA radio	Turn on battery-powered radio (get from top pantry shelf), monitor AM/FM/NOAA alerts.

In the following plan, we have a family with no out-of-state contact. They live in a rural area and Mom and Dad have CB radios in their vehicles and a base station at home.

Their son Robby doesn't have a CB radio in his car yet, but does have a cell phone, along with a backup radio: an FRS/GRMS radio in the three-day bag in his trunk (Mom wouldn't let him drive on his own without it!). Again, it is better than nothing.

Please check www.EmergencyCommunicationsBlog.com for free templates, digital copies of the examples I've shown here, and a variety of other useful information.

Plan A	Method	Cell phone
	Who	Dad, Mom, Robby
	When	As soon as possible, after emergency happens
	Phone number	Mom's cell: 426-555-1212, Dad's cell: 426-555-1213, Robby's cell: 426-555-1214
	Frequency	NA

Plan A (cont.)	Notes	If no answer, leave a voicemail, followed by text message. Then go immediately to Plan B. If phone is below 50% charge, shut it off and turn it back on for ten minutes on the hour, every hour (e.g., 3:00-3:10). If no signal is available after the first five minutes, turn the phone off and try again the next hour. If there is only one bar of battery left, only turn the phone on during even hours (e.g. 2:00, 4:00, 6:00)
Plan B	Method	CB Radio
	Who	Mom, Dad
	When	Every hour, on the hour
	Channel	Emergency channel 9. If it's busy, move to channel 16. If it's busy, move to channel 29.
	Notes	NA
Plan C	Method	FRS/GMRS radio
	Who	Mom, Dad, Robby
	When	Every hour for fifteen minutes, starting on the hour
	Frequency	Channel 16, privacy code 14 (the default setting for all of our radios)
	Notes	As with the radio schedule in Plan B, fifteen minutes every hour, calling every 2–3 minutes. If sufficient backup battery power is available, other information may be learned by setting the radio to "scan" during the rest of the hour, and listening to nearby conversations. If radio(s) are less than 50% charged or batteries are scarce, call only on even hours.
Other details	NOAA radio	Mom or whoever gets home first: Turn on battery-powered radio (get from top pantry shelf), monitor AM/FM/NOAA alerts. In vehicles, monitor radios.

GLOSSARY

AC: Alternating Current, the form in which electrical power is commonly delivered to homes and businesses. For example, the electrical outlet on the wall in your house delivers AC power.

AM: Amplitude Modulation, a way to transmit information by varying the strength of the transmitted signal.

Amateur extra license: The third and most difficult-to-obtain FCC amateur radio license, which allows the holder to operate on all amateur radio frequencies in all bands.

Amateur radio: The use of certain frequencies for non-commercial purposes, which includes but is not limited to emergency communications.

ARRL: Amateur Radio Relay League, the U.S. non-profit organization representing the interests of amateur radio operators, providing educational resources and sponsoring emergency communications services throughout the country.

Band, or Frequency Band: A range of frequencies allocated for specific use, managed by the FCC.

BOB: Bug-Out-Bag, a bag full of critical gear, providing three or more days of ready, portable emergency supplies.

Calling clock: A schedule with times, frequencies/channels, participants and other information, used to coordinate radio and other communications.

CB: Citizens' Band radio. No license is required and it uses 27-MHz/

11-meter band on specific frequencies/channels.

Crystal radio: A very simple radio receiver made from few parts and which requires no power supply.

DC: Direct Current, the way power is delivered from a battery or solar panel. Radios operate on DC power, whether from batteries or a power supply that converts AC to DC.

Emcomm: Emergency Communications.

eXRS: Extreme Radio Service, marketed by TriSquare. See Chapter 9 for more information.

FCC: Federal Communications Commission, the US Government agency tasked with managing access to wired and radio communications.

FEMA: Federal Emergency Management Administration, the US Government agency tasked with coordinating disaster response when local/state resources are overwhelmed.

FM: Frequency Modulation, a way to transmit information by varying the frequency.

FRS: Family Radio Service, a low-cost, low-power, unlicensed radio system using set UHF frequencies, often paired with GMRS frequencies in the same radio.

General license: The second of the FCC amateur radio licenses, which allows the holder to operate on all HF bands in addition to the Technician license privileges.

GMRS: General Mobile Radio Service, a low-cost, relatively low-power, licensed (currently) radio system using set UHF frequencies, often paired with FRS channels in the same radio.

Go-kit: The Emcomm operator's equivalent to the BOB, containing critical radio equipment, at least minimal backup power, and other supplies that will allow the operator to set up and run a portable radio station.

Ham radio: see Amateur radio. Why "ham"? Think "hamming it up on the airwaves." Please note, ham is not an acronym, so spelling it as "HAM" only means you're shouting.

HF: High Frequency. Used to describe the frequency range used by certain radios. HF frequencies fall between 1.8 and 29.7 MHz, and that range encompasses multiple bands, from the 160-meter band on the low end to the 10-meter band on the high end.

Mag-mount: An device that allows attaching an antenna to a metal object by using a magnetic base.

Morse Code: A method for transmitting text using tones (or other non-radio methods) that can be understood by the receiver with no special equipment. Also known as "CW" (Continuous Wave) by hams. Letters, numbers, and other symbols are represented by dots and dashes, the most famous being "SOS", represented as "...--- ..."

MURS: Multi-Use Radio Service. Five channels/frequencies are available in the 2-meter band. No license is required and MURS radios can be used in the same way CB radios are used.

NOAA: National Oceanic and Atmospheric Administration is a government agency that provides weather and other warnings.

NVIS: Near-Vertical Incidence Skywave, a method for transmitting on HF frequencies that directs radio waves directly upward, resulting in effective communications in ranges up to 400 miles away from the user.

PLB: Personal Locator Beacon, also known as an emergency beacon or distress radio beacon, used to locate the holder in an emergency.

Privacy code: A tone transmitted and received between radios which act as a squelch, preventing signals without that tone from being heard on the listener's radio. Note: radios that don't use privacy codes can hear all transmissions. The privacy code is designed to filter out other transmissions (reduce noise on the user's radio), not to provide privacy in a conversation.

Propagation: The way radio waves behave when transmitted from one place on Earth to another, often affected by solar activity, atmospheric conditions, and terrain.

SAME: Specific Area Message Encoding, a system used to encode alerts. With a NOAA radio, the SAME code will ensure you only receive alerts for a specific, relevant area.

Satphone / Satellite phone: A mobile phone that connects to satellites. As with cellular phones, satphones are two-way radios, and can usually support voice, text messaging, and data transmission.

Scanner: A radio receiver which scans stored channels or frequencies very quickly, to find active radio transmissions.

Spread-spectrum, frequency-hopping: A transmission system in which the signal is transmitted on rapidly-switched frequencies using a sequence known to the transmitter and receiver, making interception and jamming difficult, and allowing for sharing bandwidth with minimal interference.

Technician license: The first of the FCC amateur radio licenses, which allows the holder to operate on UHF, VHF and limited HF frequencies.

UHF: Ultra-High Frequency. Used to describe the frequency range used by certain radios. UHF frequencies fall between 420 and 450 MHz. The UHF ham band is also known as the "440" or "70-cm." band.

Vanity call-sign: A customized call-sign the FCC allows amateur radio operators to obtain for a small fee, as long as the format adheres to certain guidelines.

VHF: Very High Frequency. Used to describe the frequency ranges used by certain radios. VHF frequencies fall between 144 and 147.99 MHz. The VHF ham band is also referred to as the "2-meter" band.

INDEX

Made in the USA
Charleston, SC
21 May 2012